U.S. Marine Corps Raid Operations, Doctrine, Tactics, Techniques, Skills and Procedures

Department of Defense

Pentagon
Publishing

MCWP 3-43.1
(Formerly FMFM 7-32)

U.S. Marine Corps

DISTRIBUTION STATEMENT A: Approved for public release; distribution is unlimited.

PCN 143 000110 00

MCCDC (C 42)
23 Dec 2002

ERRATUM

To

MCWP 3-43.1

RAID OPERATIONS

1. Change the publication's short title on inside text to read "MCWP 3-43.1."

DEPARTMENT OF THE NAVY
Headquarters United States Marine Corps
Washington, DC 20380-0001

3 December 1993

FOREWORD

1. PURPOSE

Fleet Marine Force Manual (FMFM) 7-32, *Raid Operations,* explains doctrine, tactics, techniques, and procedures (DTTP) for raid operations conducted by Marine Air-Ground Task Forces (MAGTFs). It highlights the advantages, disadvantages, and other critical factors every commander and staff member must consider during planning and execution of a raid operation.

2. SCOPE

This FMFM is intended for use by commanders and their staffs. DTTP, planning guidance, and execution principles apply to amphibious and nonamphibious raids. However, planning requirements and execution techniques are situation-dependent and vary according to each mission.

3. SUPERSESSION

None.

4. CHANGES

Recommendations for improving this manual are invited from commands as well as directly from individuals. Forward suggestions using the User Suggestion Form format to—

COMMANDING GENERAL
DOCTRINE DIVISION (C 422)
MARINE CORPS COMBAT DEVELOPMENT COMMAND
2042 BROADWAY STREET SUITE 212
QUANTICO, VIRGINIA 22134-5021

5. CERTIFICATION

Reviewed and approved this date.

BY DIRECTION OF THE COMMANDANT OF THE MARINE CORPS

C. C. KRULAK
Lieutenant General, U.S. Marine Corps
Commanding General
Marine Corps Combat Development Command
Quantico, Virginia

DISTRIBUTION: 139 000677 00

RECORD OF CHANGE

Change No.	Date of Change	Date of Entry	Organization	Signature

(reverse blank)

Raid Operations

Table of Contents

(reverse blank)

Chapter 1
Raid Design

"Nothing is so devastating as to pounce upon the enemy in the dark, smite him hip and thigh, and vanish silently into the night."

Brigadier Orde Charles Wingate
Burma, 1943

A raid is an operation, usually small scale, involving a swift penetration of hostile territory to secure information, confuse the enemy, or to destroy his installations. It ends with a planned withdrawal upon completion of the assigned mission. Raids may be conducted as separate operations or in support of other operations. Examples of separate operations include raids for psychological purposes, destroying enemy assets not susceptible to other action, harassment, to gain combat information, as spoiling attacks to keep enemy forces off balance, and to recover or rescue friendly personnel and equipment.

OBJECTIVES

Raid design must be understood by all. The commander will articulate the specific objective and end state to be achieved by the raid force.

Raids in support of larger air, land, naval, joint, or combined operations include those planned as events within a deception story or to destroy key facilities that may influence the larger operation. Raids have recognizable characteristics whether conducted as separate operations or as part of larger campaigns. The raid force may employ a variety and combination of transportation assets such as helicopters, transport aircraft, ground vehicles, and surface/subsurface watercraft to enter or exit the objective area. Raids may be conducted by aviation, artillery, infantry, reconnaissance, combat engineers, or any other group with skills and equipment suited for the mission. Raid forces depend on surprise, detailed intelligence, timeliness of mission execution, and violence of action at the objective to ensure success.

Psychological

A raid may be conducted solely for psychological reasons; e.g., to lift the morale of friendly military and civilian personnel or demoralize the enemy. This may be necessary at the outset of hostilities, after extended periods of inactivity, or after tactical or operational setbacks. Raids conducted under such circumstances help maintain an offensive mind-set within the force.

Destruction

Raids may be required to destroy targets of such importance that indisputable confirmation of destruction is essential. Targets not easily destroyed by other means may be subject to destruction by a raid force. Political concerns regarding civilian or cultural collateral damage may also necessitate a raid mission tasking. Targets for destruction may include military or industrial installations, personnel, communication and energy facilities, and transportation nodes such as rail and port facilities, bridges, and tunnels. Raids aimed at destruction may have strategic, operational, or tactical significance.

Harassment

Enemy plans and activities can be disturbed by repeated attacks and pressure. Raids provide one method of applying this pressure. Examples of enemy targets that might be considered for harassment include isolated units, patrols, fire support agencies, combat service support, and command and control nodes. Harassment of the enemy may lower the enemy's morale as he develops a feeling of helplessness in bids to prevent these attacks.

Combat Information

Raids may be conducted to collect information regarding enemy forces including his dispositions, strengths and weaknesses, movement, reaction to attack, and weapons. All raid forces, regardless of assigned mission, perform a secondary function of collecting and reporting information.

Evacuation and Recovery

A raid may be conducted as the main or supporting effort to evacuate or recover personnel and/or material. Such raids support noncom-batant evacuation operations (NEOs), tactical recovery of aircraft and personnel and in-extremis recovery.

Diversion

A raid may be conducted as a supporting effort to create a diversion or ambiguity in the enemy commander's perception of the situation in connection with strategic, operational, or tactical deception. Assignment of alternate targets is undesirable unless the object is to create a diversion. In this case, authority to engage targets of opportunity may be granted.

EXECUTION PHASES

Upon mission receipt, the raid commander and his staff process available information, submit priority intelligence requirements, and begin mission analysis. Concurrently, security and counterintelligence plans are implemented to prevent any disclosures of operations to the enemy. Mission analysis is critical to plan development and is a continuous process by which the commander refines the focus of the operation and his intent. All elements of the raid force, the command, ground combat, aviation combat, and combat service support elements, participate in concept development.

Preparation of Plans

Preliminary organization of the raid force, selecting required personnel, and a training program are formulated once the plan has tentative approval. Plans are tested through rigorous wargaming and rehearsal. Training and rehearsal sites are selected based on similarity to the objective area and operations security. Constant examination by the commander and staff of all training and rehearsals provides a basis for evaluation and modification of plans. Applicable elements of the raid force conduct training and rehearsals with the specific aircraft, aircrews, vehicles, and ships involved in the operation. Representative aircraft and/or ships do not support preparation requirements of a raid and should only be adopted when absolutely necessary. Logistic requirements are determined and equipment procured to support the specific mission of the raid. Equipment and supplies are packaged to be compatible with specific requirements of transporting aircraft or shipping. The raid force trains with the prescribed loads to test the feasibility and utility of all equipment.

Embarkation

The most important consideration for embarkation of ships,

surface craft, and/or helicopters is support of the tactical employment of the raid force. Other considerations include capabilities of specific platforms, operations security, enroute training requirements, and flexibility for contingency response.

Movement to the Objective Area

Movement to the objective area is conducted to avoid enemy detection and response. Stealth, speed, and use of deception and disguise in movement contribute to force protection and surprise. During longer transits, frequent inspections, drills, and rehearsals are conducted to prevent deterioration of perishable skills. Intelligence updates and final preparations are performed before debarkation.

Ship-to-Shore Movement

Amphibious raids take on the added requirements of ship-to-shore movement. Planning and control of movement is the same as that conducted for amphibious assaults (see NWP 22-3/FMFM 1-8, *Ship-to-Shore Movement).* However, force size, proficiency of the ship and crew, and comfort level achieved through rehearsal may make traditional control agencies unnecessary. Silent landing techniques and care to reduce visual, sonic, and electromagnetic footprints help cover the force. Prolonged station-keeping operations offshore are difficult and increase the possibility of detection. Landing craft are either cached ashore or returned to the ship.

Approach to the Raid Objective

Once ashore, elements deploy to carry out their missions. Compromise of the force or elements of the force before reaching the objective may require execution of an alternate plan or require the commander to abort the mission. Compromise, personnel and equipment losses, and unforeseen occurrences may create a situation that prevents mission accomplishment. The commander's abort criteria are established during planning to address these and other potential problems. Movement to the objective is planned to ensure the force reaches the objective in the manner best supporting actions in the objective area.

Assault of the Raid Objective

Synchronization of the assault with execution of supporting and supported operations may require a specific timetable for attack of the target. Likewise, scheduled fires, on-station capability of attack and transport aircraft, and astronomical parameters may dictate the timing of the attack. The attack on the objective is characterized by speed and violence of action. When supporting arms are employed, they normally commence with the assault of the objective and continue through the final withdrawal. When necessary, fire support may assist the deception effort by engaging a wide variety of random targets. Actions at the objective are planned in as great a detail as intelligence and time for rehearsal permit.

Withdrawal

Withdrawal must be swift and orderly. It is influenced by the time it takes to attack and assault the objective, enemy reaction, time needed to care for and evacuate casualties, and the mode of transportation. Every effort is made to leave nothing of intelligence value. When equipment cannot be reembarked, it is destroyed. Destruction techniques for all types of equipment employed by the raid force should be made a part of individual and unit training. Evacuation of casualties is expedited.

Reembarkation

The means for evacuation must be available for immediate loading when the raid force arrives at the beach or landing zone (LZ). This is critical as the enemy may be actively pursuing the raid force. Cover is required to prevent enemy interference.

Recovery of Personnel

Contingency recovery plans are designated in the raid plan. Should elements or individuals become separated, an escape and evasion plan will be executed. Normally, recovery is attempted at 12- or 24-hour intervals for as many days as are practical without endangering the whole force. Personnel are briefed as to evasion, escape, and survival procedures beforehand and may have to depend on these skills to return to friendly areas. Immediately upon recovery, the raid force is interrogated and debriefed by designated commanders and staff.

INTELLIGENCE

Detailed planning for a raid requires precise intelligence. The availability or lack of intelligence affects all aspects of the raid operation. Good intelligence allows the commander to more precisely tailor the force size, conduct rehearsals, identify critical targets, and plan support. The Son Tay raid of 21 November 1970 points to the importance of accurate and timely intelligence. It makes clear to intelligence providers and supported commanders that information generated to support any operation is perishable and requires constant and immediate update, even while the raid force is enroute to the objective area. The raid force must be kept informed of the enemy situation and weather in the objective area to prevent being surprised.

American forces conducted a raid to liberate American prisoners of war at the Son Tay prison camp located 20 miles west of Hanoi. Meticulous detail and extraordinary effort went into planning, selecting commanders and personnel, training and rehearsal, and operations security. Intelligence support was extensive and included national and theater assets.

Execution of the raid went as planned. The force entered North Vietnam undetected. The timing of complex supporting operations was remarkably accurate. Actions in the objective area were executed as planned. Unfortunately, the prisoners had been moved to another camp and the raiders returned empty-handed.

Intelligence was so detailed that it included prisoner's names. It allowed penetration of a complex air defense system and precise employment of assault teams within the prison. Despite this level of detail, on-scene commanders did not know the prisoners were gone, even though indications were known at high levels within the intelligence community.

The raid produced a number of positive reactions by the North Vietnamese toward their American prisoners and raised the prisoner's morale. However, the purpose of the raid--to free American prisoners of war--was unfulfilled.

TRAINING

Commanders must ensure that training the raid force involves all participants. Compartmentalized training, insufficient individual training, and failure to conduct thorough unit training create the conditions for failure. Well-planned and aggressive training will prepare the individual Marine, uncover flaws in the plan, match capabilities of elements to assigned tasks, allow the employment of a smaller force, and provide situational awareness to all raid force members. Thorough training of the raid force fosters initiative and confidence to overcome unforeseen obstacles. The attack of Fort Eben Emael by the Germans at the outset of the 1940 western offensive illustrates the value of meticulous training.

The German offensive begun on 11 May 1940 ended the "phoney war" on the western front. In addition to seizing the Netherlands, Army Group B was tasked with making the Allies believe the attack through northern Belgium was the main attack. A key element to the plan was the gliderborne attack of the "impregnable" fortress, Fort Eben Emael and key bridges along the Albert Canal. The fort, considered to be the strongest in Europe and manned by 800-1,200 Belgian troops, was the objective assigned to Assault Force Granite (75 troops and 11 glider pilots noted for their individualism and fearlessness). As mission success was critical to the Army Group's rapid movement into Belgium, no effort was spared in preparing and training the assault force.

Assault Force Granite was provided over 300 miles of the Benes line fortifications, located along the Czechoslovakian border, for casemate assault training. Training ranged from how to get in and out of gliders to piloting them at night in formation, combat loaded, and landing with pinpoint accuracy. Assault troops were instructed in the use of new weapons, most notably the 110 lb shape charge, that were critical to destroying key parts of the fort. Airplanes and gliders were matched to ensure the pilots, familiar with each other through extensive training, could anticipate the other's actions. Troops studied terrain models, photos, and table models to ensure complete knowledge of the objective area. Glider pilots took part in ground training as they became an integral part of the assault force after their piloting duties were over. After months of training, the assault force knew every detail of the fort and the overall plan. A Corporal Alefs, reflecting on the training, mentioned that the 85 troops felt stronger than the 1,200 they were to attack.

The operation started with a harbinger of disaster. The attack commenced without the commander when his glider was released before reaching proper altitude. A sergeant assumed command as the nearest officer was involved in another action and could not be reached. However, in spite of these serious problems, the operation was a resounding success. The sergeant who took command later remarked that "the officers had trained all of the men so well that the officers were expendable."

SELECTION OF OBJECTIVE

Raids are inherently dangerous. The raid force may be completely destroyed. Targets must be closely scrutinized. Commanders must ensure that raid objectives are worth the risk, leading to strategic, operational, or tactical gain. The attack of the Lofoten Islands is an excellent example of an ideal selection. It met a military need; the location and enemy defenses made it susceptible to attack; it was within the capability of a well-trained but inexperienced raid force; and it provided a much needed morale boost to the nation.

The first 2 years of World War II witnessed one Allied setback after another. The British, limited in resources and alone, continued the fight through raid operations against German military and economic targets along the Atlantic coastline. On 4 March 1941, Numbers 3 and 4 Commando and 52 Royal Engineers struck the Lofoten Islands, located off the northern Norwegian coast. (See page 1-10.)

The Lofotens provided the Germans the majority of their fish oils, an essential ingredient in manufacturing explosives. The nearest German airfields were iced over at this time of year and the nearest garrison able to influence the battle was almost 60 miles away. It appeared resistance from both ground and air forces outside the objective area would be slight. Numerous raids had been planned and cancelled, frustrating the commandos who had been well-trained for these operations. These factors all mitigated in favor of the Lofotens attack.

The Lofotens raid went extremely well. Fish oil factories were destroyed, a number of German prisoners taken, and over 300 Norwegians returned with the commandos to serve with the Allies throughout the remainder of the war. The commandos suffered only one casualty. The raid was filmed and shown throughout Britain to the delight of every audience.

SURPRISE

Surprise is a force multiplier essential to raid operations, and is achieved by many means. The capability to conduct the unexpected raid creates for the enemy a host of contingencies for which he must prepare. The raid force's ability to operate over unexpected routes and terrain which is generally considered impassable and in conditions of reduced visibility enhances the probability of attaining surprise. Deception and operations security also contribute to surprise. Throughout the operation, the commander can maximize the use of ruses, feints, and false insertions to confuse and disrupt the enemy. Conventional operations are also used to mask the raid operation.

LOFOTENS ISLANDS RAID, 1941

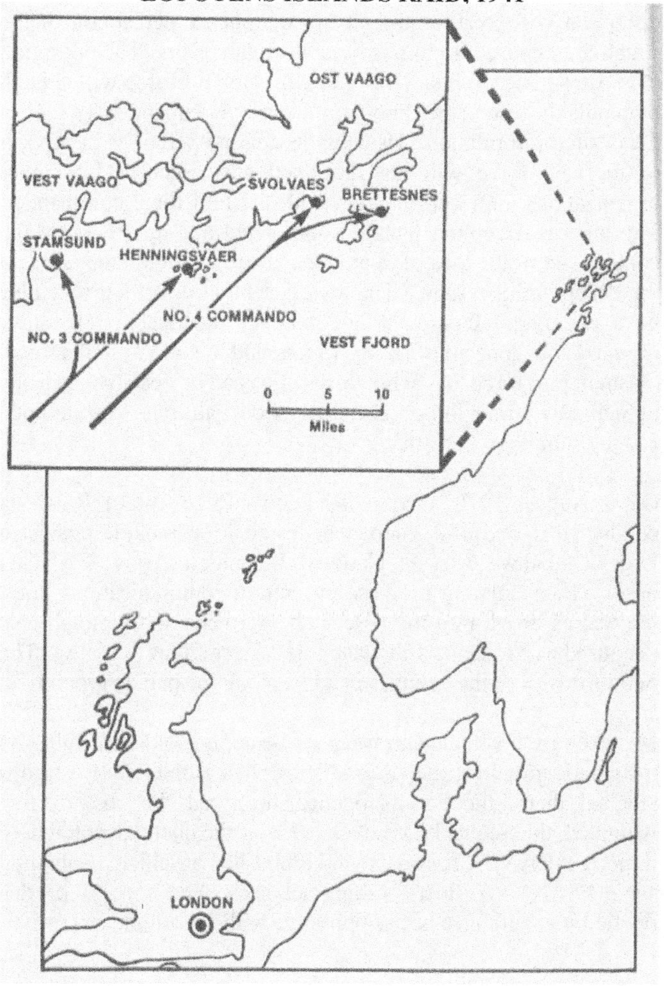

Surprise can be achieved through deception. Successful deception will create ambiguity in the enemy perception of the situation or cause him to misdirect his main effort. The deception plan, story, and events orchestrated by the raid force will depend on available resources, time, and the enemy capability to collect the desired information. Usually, deception operations in support of the raid force will be developed and executed by higher organizations after coordination with the raid force commander. The tactical deception plan is coordinated and approved by the commander of the area of operations affected by both the raid and the deception operation. The area deception operation may also be a deciding factor as to the time of the attack, force task organization, concept of the operation, and method of withdrawal. A small raid force that achieves surprise can be decisive against a numerically larger force. A Selous Scout raid into Mozambique is one example of this affect.

On 9 August 1976, the Selous Scouts Regiment of Rhodesia conducted a raid 100 kilometers inside Mozambique against a large Zimbabwe African National Liberation Army (ZANLA) base. After isolating the base by cutting communications lines, the Scouts drove into the base with 14 trucks and armored cars disguised as Frente da Liberataco de Mocambique vehicles. The Scouts deployed the vehicles in a line along the parade ground.

Mistaken for a vehicle and weapons resupply, soldiers drilling on the parade ground rushed the vehicles. Just as the ZANLA troops reached them, the Scouts opened fire. At the cost of five wounded, the Scouts killed over 1,000 of the enemy, captured 14 important ZANLA terrorists, and ambushed a vehicle containing senior ZANLA officers, killing all six. The success of this operation would have been impossible without complete surprise.

WITHDRAWAL

The raid force, even after a successful attack, is most vulnerable after actions in the objective area. The enemy, now alerted to the presence of the raid force, initiates actions to reestablish contact with or interdict the raid force during its withdrawal. The raid force may now have casualties, evacuees, and captured personnel and/or equipment that further complicates its withdrawal. Planning the withdrawal requires the same forethought as any other event in the raid operation. A 1945 raid to liberate American and Filipino prisoners of war serves as a good example of the complexity and importance of withdrawal planning.

On 30 January, 121 men of the 6th Ranger Battalion, assisted by 90 Filipino guerrillas and 160 Filipinos that served as porters, conducted the raid at Cabanatuan in the Philippines.

CABANATUAN, 1945

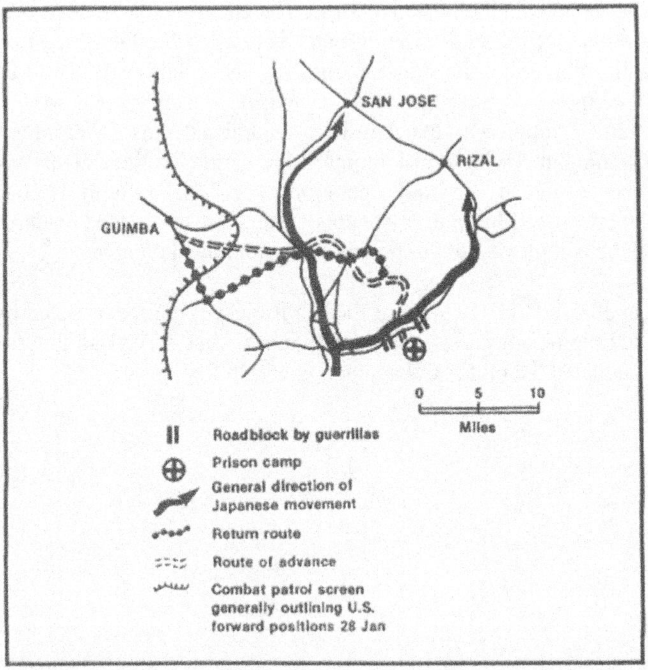

In a well-orchestrated attack, the Rangers assaulted the camp, rescued 513 prisoners, and began the withdrawal at 2015, 30 minutes after the action was initiated. The withdrawal required a movement of 25 miles, under pressure from an estimated 2,000 Japanese soldiers. Many of the 513 prisoners were in terrible physical condition. However, the raid force came equipped with 51 carts to transport those prisoners incapable of walking, enabling the raid force to reach friendly lines at 0800 the next day. The withdrawal, under these more than difficult conditions, would not have been possible without the superb planning involved. This action demonstrates the importance of withdrawal planning.

MOBILITY

Obviously, to attack an objective, the raid force must be able to reach it. Commanders may possess knowledge of critical targets, yet not be able to attack if their force lacks the requisite mobility. The raid force commander may have a variety of mobility assets in which the raid force can reach the objective. The raid force commander should consider all of these assets individually and in combination for the most suitable means. The importance of mobility and the imaginative combination of these assets to allow attack of the target is best illustrated by the 3 July 1976 Israeli long-range international raid to rescue 105 hostages held by terrorists at the Entebbe International Airport, Uganda.

The incident began a week earlier when Palestinian terrorists hijacked an Air France aircraft, originating from Ben-Gurion Airport, shortly after an intermediate stop in Athens, Greece. The flight terminated at Entebbe, 2,187 miles from Tel Aviv. The great distance, coupled with the apparent support of the Ugandan dictator, complicated any attempt to rescue the hostages through military action. In fact, most
in the government had reluctantly conceded the need to negotiate a terrorist-hostage exchange to save the passengers' lives. Concurrent with negotiations, the leadership continued to look at a variety of military options.

The plan for Operation Thunderball (the codename for the Entebbe raid) evolved over a 48-hour period. Insertion and extraction of the force--one of the most time-consuming and difficult actions--required extraordinary operations security, coordination, and logistics planning. The plan involved the synchronization of four C-130s and two disguised Boeing airliners, and establishing a return-trip refueling site in Kenya. The terrorists, secure in the knowledge that the separation between Uganda and Israel made a rescue attempt remote, were taken by complete surprise. The Israeli paratroopers landed at Entebbe at 2303, assaulted the terminal, killed every terrorist and approximately 20 Ugandan soldiers, and rescued the hostages at a cost of 1 paratrooper and 3 hostages killed. The raid was a strategic windfall, reinforcing the perception that Israel would not tolerate terror against its citizens regardless of where it occurs. Israeli prestige and confidence could not have been higher.

SIZE, COMPOSITION, AND TASK ORGANIZATION

The smaller the force, the better. This requires exhaustive cross-training of all members to ensure required capabilities are not lost due to attrition of the force. Smaller forces have numerous advantages over larger formations. The reduced requirement for lift enhances stealth. The small footprint ashore becomes more difficult for the enemy to target and otherwise open terrain may conceal a smaller unit. The urge to employ a larger unit for security's sake may contribute more toward compromise and undesirable enemy contact than any true security a larger force may enjoy. *Quality over quantity* as well as capability are the keys to determine force size.

Size, composition, and task organization are important because they directly influence many other factors such as speed, surprise, and logistics. Task organization is ordinarily associated with combining units to accomplish the mission. Small forces, such as raid forces, are equally concerned with the composition of the force, the combining of individuals or skills necessary to accomplish the mission.

A key to a successful raid is the ability to complete objective area actions before the enemy has time to react. Men and equipment are evaluated in terms of overall contribution. The British Special Air Service (SAS) Regiment, formed in November 1941 and operating in small detachments, conducted a variety of long-range raids and sabotage missions against German and Italian forces. The basis for organizing in small teams came from the realization that a four- to five-man detachment could infiltrate to the objective and inflict significant destruction aided by the element of surprise whereas a larger force would have to fight to the target.

SAS operations were conducted across the vast expanse of desert extending throughout North Africa. This environment was harsh and difficult to navigate. The weather was unpredictable. The SAS turned to the Long-Range Desert Group (LRDG)--a unit with extensive experience in desert reconnaissance--for assistance.

On the night of 12 December 1941, an SAS detachment, transported by a LRDG patrol, conducted a raid against the Axis airfield at Tamet. Twenty-four aircraft were destroyed without loss to the raid force. This task organization of SAS and LRDG elements resulted in a combina-tion of capabilities that would eventually destroy over 300 Axis aircraft. A captured German officer would later remark to a LRDG patrol leader "We Germans couldn't do this sort of thing-out 500 hundred miles from our base for days or weeks on end. We like to go about in a crowd."

COMMUNICATIONS

Communications is part of the information exchange system that supports and controls the flow of information throughout the raid force. It is the dissemination of information to those key individuals, elements, units, or the entire force. Effective communications are required to ensure every member of the raid force understands the situation and the actions that must be taken to adapt to changes to it. Many military operations have failed due to the inability of the commander to redirect his subordinates once he has recognized the need to change the plan. An example that epitomizes the detrimental effect of poor communications planning is the failed Iranian hostage rescue attempt in 1980.

A decade after the Son Tay raid, another raid to rescue American prisoners was undertaken. On 24 April 1980, a task force consisting of Special Forces, Ranger, Marine, Air Force, and Navy personnel attempted to penetrate Iranian territory, assault the United States embassy, and repatriate the 53 hostages taken when Iranians overran the embassy 6 months earlier.

The operation began with the launching of eight RH-53D helicopters from the USS Nimitz, two more than the minimum of six felt necessary to succeed. Soon, the helicopters had penetrated Iranian territory enroute to "Desert One," a refueling and rendezvous site. One helicopter experienced mechanical problems (a rotor blade failure) and landed to inspect it. A second helicopter landed to assist and when it was decided to abandon the stricken aircraft, the crew boarded the second helicopter and resumed its flight. The seven remaining helicopters then ran into a massive dust cloud, 3 hours into the flight. The helicopters became separated, emerged into the clear, only to run into a larger and more dense dust cloud an hour later. Though planners were aware of the dust cloud problems that might be encountered, the pilots of the helicopters and C-130s had not been briefed of this phenomenon.

Another helicopter was soon lost when a cooling system failed and disabled the flight instrumentation. Hampered by poor visibility and disoriented by vertigo, the pilots returned to the Nimitz. Because of communications restrictions, the crew was not aware that one helicopter was already lost and that they would be out of the dust within a few more minutes.

The remaining six helicopters arrived at Desert One from 1 to 1-1/2 hours late. Of these, one experienced a hydraulic system failure and could be flown only with great risk. The mission commander was forced to abort the operation when the number of available helicopters fell below the abort criteria of six.

Strict radio silence had been imposed during the helicopter flight to Desert One and prevented critical operational reporting when the unexpected occurred. Silent communications such as signal lamps were authorized, but when the helicopters entered a massive dust cloud, these became useless. This lack of communications hindered the ability of the helicopters to arrive as planned and contributed to significant delay and confusion at the site.

The Holloway Review Board, investigating the failed operation, found the U.S. force did not understand the Iranian signal intelligence capability, that secure communications were available with the force while enroute, and that selected use of radio communications may have resulted in a "more favorable execution of the movement to Desert One." The board urged that planners of future operations conduct a comprehensive analysis, assessment, and training in matters of signal security operations and planning.

Another factor that contributed to the communications problems was the failure to conduct a full-dress rehearsal. The overall commander of the raid operation, for reasons of operations security, had decided that "regular integration of training was undesirable." Rehearsals were conducted with representative aircraft, not all hands, and a thorough test of the communications plan, to include contingencies such as adverse weather, was not conducted. Few peacetime military operations will match the Iranian hostage rescue attempt in interest, debate, or impact upon national pride, as the nation was glued to news of the hostages every day until they were finally released.

REHEARSALS

Rehearsals must be conducted to validate plans, ensure precise execution, correct deficiencies in equipment, personnel, and scheme of maneuver. Though the Son Tay raid force lacked the definitive intelligence that would have allowed them to succeed, they spared no effort in preparations for actions in the objective area. Two months before the raid, rehearsals began in earnest, with 3-day and 3-night rehearsals conducted each 24-hour period. These rehearsals consisted of dry runs and live fire. Locations of each man and each round was known to the extent that if an individual was off the mark even a yard, it would result in a friendly fire casualty. Assessments were conducted after each rehearsal.

Alternate plans were also rehearsed. Each plan was rehearsed to the same standard, during day and night, increasing in tempo and vio- lence until the commander was satisfied. In fact, an alternate plan was executed at Son Tay when one helicopter landed and discharged its passengers at the wrong complex 400 meters from the prison. The subordinate commanders executed the alternate plan faultlessly.

During a briefing for Secretary of State Henry Kissinger before the raid, the last question he asked was "What if the raid fails?" The response impressed him. He was told the raid stood a 95 to 97 percent chance of success as air crews had flown 697 hours in 268 sorties and the operation had been rehearsed 170 times.

UNITY OF COMMAND

In the course of the operation, the raid force will face a number of decision points that must be reconciled without delay. Go/no-go decisions before launching the attack, abort decisions during movement to the objective, and decisions to execute alternate plans all require quick action on the part of the raid force commander. These decisionmaking requirements mandate the maintenance of a clear, unencumbered chain of command and the authority of a single commander.

During World War II, the British created Combined Operations to supervise the conduct of all raiding activities against the European continent. Combined Operations, headed by Lord Mountbatten, reviewed, approved, and tasked operational units with the conduct of raids, and served as the conduit for all raiding requirements. One raid in particular, the Bruneval Raid, illustrates the value of unity of command in what could be a complex, multiservice operation. The raid was conducted by soldiers, supported by the Royal Navy and Air Force, for the benefit of the British air effort against Germany.

The technological battle of the war included the efforts to produce superior radars for both defense and direction of offensive air operations. As German night fighter and bombing capabilities seemed to improve beyond that resident in the British forces, it was concluded by the Air Ministry and British scientists that these capabilities benefited from German radar and radio direction enhancements. A proposal to conduct a recovery raid against a radar site near Bruneval, France was submitted by the Air Ministry to Combined Operations. The object of the raid was to recover components of a *Wurzburg* radar
system for study. Combined Operations approved the concept and tasked one company of paratroopers, with a detachment of Royal Engineers and radar technicians to conduct the raid. (See page 1-20.)

Intelligence from aerial photography and the French underground provided enough detail to allow for precise planning. The raid force, under the command of Major John Frost, trained and operated with near complete autonomy, supported by Combined Operations, which coordinated naval and aviation support for the raid force. The raid force inserted by parachute on the night of 27 February, 1942, accomplished its mission, and withdrew under pressure to waiting assault craft on the beach. Major Frost retained complete authority throughout the raid. He supervised the attack on the objective, security, and execution of the withdrawal, without the prospect of higher authority intervention. At the higher level, the authority of Combined Operations to organize and task forces to mount raids provided the unity of command at the strategic level, mitigating against interservice or branch interference.

SIMPLICITY

The more complicated the plan, the greater the potential for failure. Since raid forces are as susceptible to the friction of combat as any other force, the raid force commander must strive to keep his plan as simple as possible, detailed, and clearly understood by his subordinates. Detailed plans should not be confused with complicated plans. As the tasks planned to accomplish each critical event become more compli-cated, the risk of failure increases on each successive component of the task. One failure can unravel the entire plan and force the adoption of an alternate plan or mission abort. Simplicity was key to Confederate raid operations against Union forces in Northern Virginia.

Colonel John S. Mosby of Civil War fame, was noted for his successful antagonism of Union forces throughout Northern Virginia. He conducted numerous raids against isolated Federal forces in the midst of powerful Union formations. His raids, noted primarily for their audacity, were also surprisingly simple. One raid in particular, known as "The Greenback Raid" for the payroll seized by his raiders, exemplify the fundamental of simplicity.

BRUNEVAL RAID, 1942

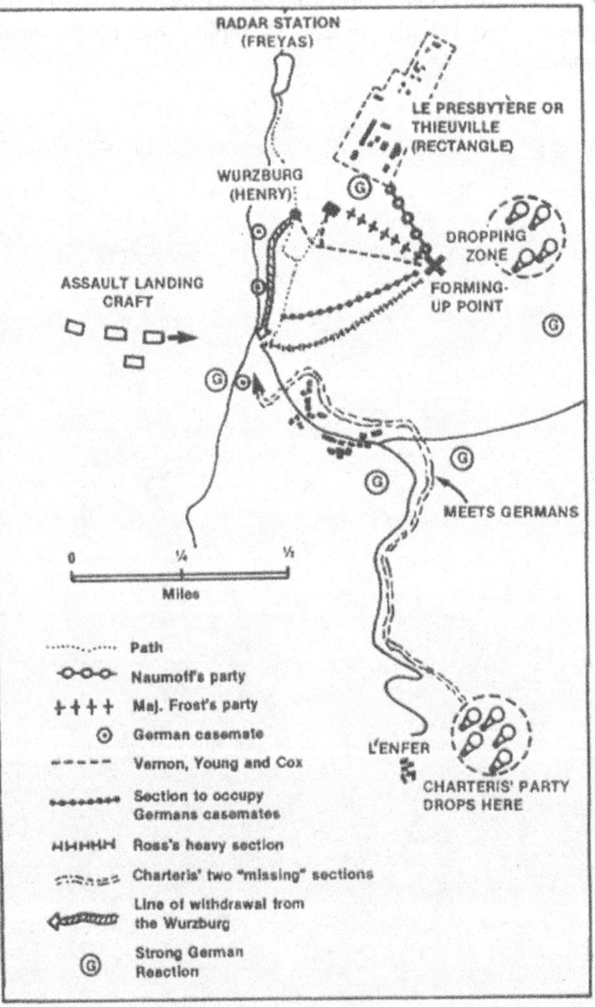

Colonel Mosby, acting on intelligence revealing a gap in Union dispositions, conducted a destruction raid on a west-bound passenger train the night of 13 October 1864 between Martinsburg and Harper's Ferry. The purpose of the raid was to force Union General Philip H. Sheridan to divert more combat power to protect his lines of communications. The plan was simple; infiltrate through the gap in Union forces, sabotage the rail and derail the train, then burn the rail cars to sensationalize the act. The plan was executed without incident, so well in fact that Mosby and the majority of his men were asleep when the train derailed with a tremendous crash. Wakened by the ensuing commotion, Mosby's raiders methodically looted and torched the rail cars, and quickly made their escape unscathed with horses, prisoners, and over $170,000 in payroll.

The effects of the raid were achieved in short order. Union forces were quickly posted along the railroad at the urging of General Stevenson of Harper's Ferry (who requested a minimum of 1,000 cavalry troops from Sheridan for the task). Force redeployments for security were not limited to the railroad. The raid also reinforced the inaccurate perception of significant Confederate strength and the need for increased security forces throughout the region. This and many other raids conducted by Mosby, bold but simple in design, drained resources from the Union throughout the war. His ability to mount almost continuous raiding activities can be traced to his adoption of standard procedures for movement, rendezvous, and simplicity in objective area actions.

Chapter 2
Organization of the Raid Force

A raid force is formed by task-organizing and training a raid force from existing force elements. The nucleus of most raid forces is the ground combat element (GCE). Its contribution may range in size from squad to battalion. Specific organization depends on mission, enemy, terrain and weather, troops and support available—time available (METT-T). The raid force is normally organized into functional groups, each specifically tailored to carry out essential tasks. A raid force may consist of raid groups, elements, or teams. In this respect, a raid force may consist of a command element, a reconnaissance element, a support element, an assault element, a security element, and a reserve as shown below.

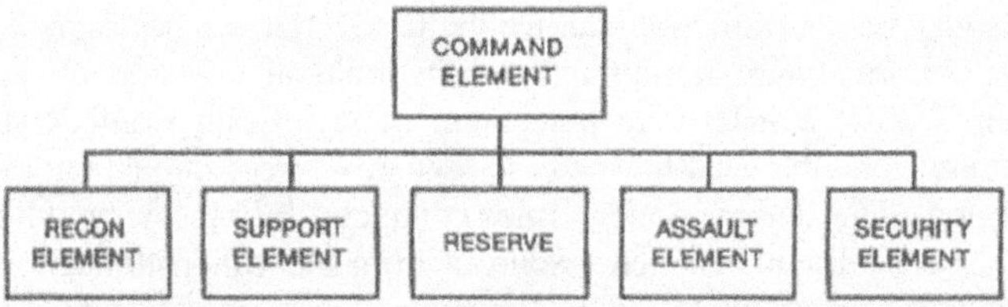

COMMAND ELEMENT

The command element controls movement to and actions at the objective and the withdrawal. It consists of the commander, and when designated, the assistant commander, and the facilities required to command and control the raid operation. Organizational staffs may provide requisite support to the raid force prior to execution yet not be part of the command element. When the raid force is part of a larger organization conducting extensive operations on a continuing basis (such as a rifle company is to an infantry battalion), detailed plans may be developed and coordinated at a higher level and executed without a staff.

RECONNAISSANCE ELEMENT

The mission of the reconnaissance element is to provide continuous observation of the objective, last-minute intelligence to the commander, and initial security for the raid force. Organization of this element is determined by the mission, size and type of enemy force, enemy mobil-ity and state of alert, terrain and avenues of approach, and the time needed to isolate the objective area. The reconnaissance element may be tasked with specific security tasks during movement, actions in the objective area, and withdrawal. Tasks may include pre-H- and L-hour reconnaissance and security of landing points or LZs, establishment of observation posts, initial terminal guidance, and target surveillance.

SUPPORT ELEMENT

The support element provides the heavy volume of fire needed to neutralize the objective. Because fires from this unit are violent and devastating, they must be closely controlled to ensure the precision required. On order or as planned, fires are lifted and shifted to cover the maneuver of the assault element by suppressing enemy fire. The support element may also be given specific locations to cover by fire in support of the security element if an enemy quick-reaction force moves toward the objective area. These may include routes into and out of the objective site, key terrain features, or installations adjacent to the main objective. Once the assault has been completed or on order from the raid force commander, the support element covers the withdrawal of the assault element or displaces to the next planned position. Support element organization is determined by--

- The objective's size, the surrounding area's geography, and the area's enemy threat (including air).
- The assault element's mission.
- Suitable firing positions.
- The size and nature of the enemy force in the objective area and those enemy forces capable of reinforcement at the objective.
- Fire support from other units (aviation, naval surface fire support, and artillery fire).

ASSAULT ELEMENT

The assault element is normally tasked to conduct the preponderance of objective area actions; i.e., accomplish the mission. Normally, it is inserted after the reconnaissance, security, and support elements and is the first element to withdraw and reembark.

METT-T considerations determine the organization of the assault element. The assault element may be organized into one or more assault teams, prisoner teams, search teams, medical teams, demolition teams, or breach teams.

To destroy a point target or installation in a heavily defended area when aircraft cannot get close enough to be effective, the assault element may consist of no more than a team equipped with laser target designators which could then be used to direct air-delivered laser-guided munitions from a safe distance.

SECURITY ELEMENT

The security element inserts after the reconnaissance element and covers the advance of the assault element to the target and its subsequent withdrawal. To succeed, the security element must orient on those enemy forces capable of influencing the operation within a given timeframe. Frequently, security elements may be required to establish blocking positions along lines of communications to interdict and delay enemy forces reacting to the attack. Other tasks that may be assigned include--

- Securing the objective rally point.
- Providing early warning of enemy approach.
- Preventing enemy escape from the objective.
- Providing overwatch for the units at the objective and suppressive fires for their withdrawal.
- Providing short-range air defense.
- Providing initial terminal guidance.

RESERVE

The raid force is task-organized to execute its assigned mission without a reserve. A true reserve may or may not be designated by the raid force commander. This decision is based on the ambiguity of the situation, mobility assets available, the affect on surprise and speed, and the combat power required to accomplish the mission. An alternative to designating a reserve is to assign subordinate elements responsibility for conducting those preparations necessary to accomplish contingency tasks. The inability of the subordinate element to accomplish its primary task and a contingency task simultaneously may necessitate designating a true reserve. A reserve may be retained afloat or in rear areas and inserted only if the raid force needs assistance in accomplishing the mission. When employed, the reserve of a raid force is similar to a reserve in other ground combat operations. However, when a mission is particularly hazardous or vital, a reserve may be inserted and employed with the raid force.

COMMAND RELATIONSHIPS

Experience during World War II and the Korean War of both U.S. and British forces clearly shows the advantages of controlling and directing all raids from the highest command in the area of operations. This does not imply that subordinate commands cannot or do not plan and conduct these raids, but area commanders are kept informed and, in effect, approve such raids to ensure unity of effort and coordination within the area of operations. Such control is necessary to avoid duplication and conflict between units which might interfere with each other. The principles of organization and of command relationships applicable to all operations apply to the raid. Due to the unique nature of raid operations and the resulting variations in troop and naval forces, it is beyond the scope of this manual to outline the command relationships of every conceivable situation.

Overall command of the raid rests with the commander of the area of operations affected. The commander issuing the initiating directive specifies the exact details of command relationships. *The raid force commander must report only to one senior.* The immediate commander of the raid force commander provides the raid force commander with all required support and coordinates with outside agencies and higher echelons. The superior *must* specify conditions under which basic plans may be changed, the raid delayed or aborted, and similar contingencies. When extensive raid operations are required over an extensive period of time to satisfy a mission need, a provisional organization may be established to provide for overall control, planning, and execution of raid operations.

Commander, Amphibious Task Force

The commander, amphibious task force (CATF) establishes overall command and control procedures for the amphibious raid. He may delegate his command authority over a raid force commander to a subordinate Navy commander; e.g., that of an advance force with a corresponding raid force. Navy staff planning establishes procedures for embarkation, movement to the objective area, landing, fire support, and the withdrawal, after consultation with the raid force commander. It also covers composition and action of the naval escort and usually gives a rendezvous at sea within striking distance of the coast.

Planning lays down arrangements for signals and communications between ships and shore and establishes recovery procedures for picking up raid force personnel who may be stranded ashore after reembarkation is completed. Raids originating ashore and conducted against land targets may be supported by any number and type of ship, craft, and special detachments. *Although CATF and commander, land-ing force (CLF) are still used in Joint Pub 3-02, Joint Doctrine for Amphibious Operations, emerging doctrine of the naval expeditionary force may or may not retain these terms. Nevertheless, new command relationships with new names will function very similarly to CATF and CLF.*

Marine Air-Ground Task Force Commander

Marine Air-Ground Task Force (MAGTF) planning is primarily focused on integrating all subordinate element actions in support of the raid force. Raids in support of sustained land combat are conducted through the authority inherent in the established chain of command. The MAGTF, GCE, and aviation combat element (ACE) commanders may conduct raids that support their respective schemes of maneuver. Raids requiring participation of all MAGTF elements will normally originate from the MAGTF commander. He may direct all subor-dinate elements to provide one particular element the necessary support required to conduct a raid. For example, a raid originating from and in support of GCE operations may receive assault support from the ACE as directed by the MAGTF commander.

Aviation support, in particular, may be complex and require significant planning. Aviation elements supporting the raid may be shore- or carrier-based. Pros and cons of an air attack before inserting a raid force are carefully balanced. Although a well-timed air attack inflicts casualties and damage and drives the enemy forces underground, it also increases his alert status. Planning is focused toward support of employment and withdrawal of the force, attack of selected targets, and contingency plans. Plans for air attacks are closely timed and coordinated with ground maneuver, naval surface fire support, and artillery support planners to ensure adequate and integrated fire support is available throughout the operation. Aircraft can make a contribution as part of a tactical deception plan by striking targets throughout the enemy's depth, conducting false insertions, and conducting conditioning flights for several days or nights before the operation. The enemy may become accustomed to regular air activity and fail to associate this activity with a specific raid target.

Raid Force Commander

Subject to the overall authority of the CATF, MAGTF commander, or attack group commander, as appropriate, full responsibility for the conduct of operations ashore is vested in the raid force commander. The planning and, in most cases, the execution of the landing, assault, and withdrawal are primarily his concern. The feasibility of raid force proposals is considered from the standpoint of the effect on the overall operation. Proposed raid force courses of action that make demands beyond the naval force's capabilities will be rejected. The raid force commander and his staff will develop the raid concept of operation. Raid force planning requires the latest and best information available as to actual conditions in the objective and target areas. Likewise, planning techniques employed by the raid force must expedite execution to maximize the relevance of that intelligence. As late intelligence arrives, changes are made as necessary and coordinated with parallel Navy, MAGTF, and raid force planning staffs. Integration of all supporting plans and the raid concept is conducted at the raid force level. Depending on METT-T, the commander may be from an element other than the GCE. The raid concept of operations provides the commander's vision for the conduct of the raid from departure through recovery.

(reverse blank)

Chapter 3
Raid Planning

Raid planning is characterized by coordinated, thorough, and detailed planning by the raid force, supporting, and supported organizations. Parallel planning for the raid is conducted concurrently by Navy, MAGTF, and raid force staffs, as appropriate. Each staff has special concerns, but all work to the common mission of the raid force and production of the raid plan. The raid force is the supported organi-zation and should include aviation, ground combat, and combat service support staff representation. Raid operations are planned and exe-cuted in accordance with procedures delineated in Joint Pub 3-02; *Joint Doctrine for Amphibious Operations;* Joint Pub 3-02.1, *Joint Doctrine for Landing Force Operations;* and FMFM 3-1, *Command and Staff Action.*

PLANNING SEQUENCE

Critical events within the conduct of a raid are embarkation, movement to the objective area, ship-to-shore movement, movement to the objec-tive area, objective area actions, and withdrawal. Though these actions are executed in sequence, the *arrangement of events for planning is different.* There are sound reasons for this variation. The preeminent event is the attack of the target. All planning to support other events is designed to facilitate objective area actions. Force size necessary to accomplish these actions establishes the basic requirements for logistics support. The nature of the target and the enemy capability to disrupt the attack establishes fire support requirements. The direction of attack may establish the landing, departure, or insertion location. *Actions at the objective area form the basis for all other planning.* Any part of the plan which fails to support the attack is rejected.

RECEIPT OF MISSION

The receipt of the mission is the event that initiates the raid force commander's planning process. The mission may be received via a number of means (written orders, oral orders, electronic means, graphic depiction, or a combination). Missions may also be deduced by a commander based on the higher commander's intent or the outcome of a current operation. Regardless of how the mission is received, a commander initiates the planning process for the mission at hand and alerts his subordinates by issuing a warning order. This initial warning order may lack specifics, but provides subordinates knowledge of a pending operation. Warning orders are issued as required throughout the process to update subordinates sufficiently to prepare for the operation.

MISSION ANALYSIS

The raid force commander's mission analysis is the first and perhaps the most important step in the planning process. It identifies for the commander and staff the salient tasks that must be accomplished to accomplish the assigned mission. Mission analysis is a continuous process, the products of which may be altered based on new intelligence or a changed tactical situation. Mission analysis consists of command and staff actions related to determining information requirements through the gathering of facts and making assumptions, analysis of the higher organization's mission and intent, and determining the unit's mission and required tasks.

Information Requirements

The raid force commander determines the information requirements on the area of operations, the relative combat power of friendly and enemy forces, and on enemy capabilities. That which is not available must be collected and integrated into the estimate process.

The raid force staff determines facts on each staff officer's own functional area. These provide the raid force commander an accurate understanding of the current situation from which he may deduce assumptions and assessments of capabilities and operational factors.

The raid force commander receives information available from his staff and from higher, lower, and adjacent headquarters. Gaps in information at this early stage are covered by making assumptions. Assumptions replace necessary but missing facts. An assumption is appropriate if it meets both tests of validity and necessity. However, the raid force commander should avoid an over reliance on assump-tions as this tends to increase risk to the force. Validity defines whether or not an assumption is likely to occur. "Assuming away" potential problems is an example of using an invalid assumption. Necessity defines whether or not the assumption is absolutely essential for the development of a plan. If planning can continue without making the assumption, the assumption fails the test of necessity and is consequently inappropriate. Staff officers must continuously reevaluate assumptions as part of the estimate process. Additional facts should replace assumptions as early as possible to reduce uncertainty in the execution of the plan.

Identification of Tasks

Specified tasks are stated in the higher headquarters order or plan. They come primarily from paragraphs 2 and 3 but may be found elsewhere, like in coordinating instructions or annexes. Any task that pertains to any element of the unit should be identified and recorded.

Implied tasks are those tasks not specifically stated in the order or plan that must be accomplished to satisfy the overall mission or to satisfy any of the specified tasks. Implied tasks come from further analysis of the order and analysis in conjunction with the known enemy situation and the terrain. Inherent, routine, or standing operating procedure (SOP) tasks are not included in the list of tasks.

Although the raid force commander is responsible for the identification of the essential tasks and the mission, the staff develops a tentative list of essential tasks and a preliminary restated mission. From the list of specified and implied tasks identified earlier, those tasks that define the success of the mission are identified as *essential.* To properly identify essential tasks, the staff and commander must be aware of the intent of the commander ordering the mission and that of his superior.

Identification of Limitations

Limitations are restrictions placed on a commander specifying things that cannot be done and/or things that must be done. In most cases, limitations prescribe some detail of raid force organization, maneuver, or rules of engagement.

Identification of Available Assets

The staff considers assets allocated in the task organization, attachments and detachments, or in organizations for combat in the execution paragraph. More importantly, the relationship between mission and assets is critical. The folding together of time, space, and assets is critical to mission success.

Acceptable Risk Levels

The higher headquarters might specify a risk the tasking commander is willing to accept to accomplish the mission or provide guidance pertaining to the use of friendly capabilities.

The level of acceptable risk can be deduced from analysis of mission and intent.

Initial Time Analysis

The element of time is not clearly identified for analysis in any of the staff estimates. However, time analysis must be an integral part of the mission analysis, and must be conducted continuously until the mission is accomplished. Time analysis uses the following guidelines:

Determine Time Available. Time available begins with receipt of the mission and ends with the time of execution and completion of the mission stated in the order.

Know the Decision Cycle. A decision cycle is the total time required from the time the commander obtains information, processes it, makes a decision, issues orders, and subordinates execute the operation. This is a necessary tool to properly plan time allocation.

Time Allocation. Once a raid force commander has a general under-standing of his mission and the time available, he must allocate the time for the various phases of the operation. This allocation is often determined by reverse planning. Knowing the time to execute the operation, the raid force commander must consider how long the unit needs to move to the point of execution; time for rehearsal, and time for planning. Since subordinate units also require planning time, the amount of time available for their planning must also be allocated. The commander should consider the 1/3 - 2/3 rule. Each headquarters should use roughly one-third of the planning time available and allow two-thirds of the time for subordinate units and elements. Each element should use only the minimum amount of time necessary to plan an operation. Time allocation must consider the decision cycle of the unit doing the planning and that of the enemy.

Restated Mission

The restated mission will be the mission statement for the raid force. It contains all the elements of a mission statement (who, what, where, when, and why). "What" is the essential task(s), the accomplishment of which defines mission success. If there are multiple tasks, they are listed in the sequence in which they are expected to occur. At this stage, the restated mission should not be changed. However, it may change if an error in analysis is discovered or the mission is changed by higher authority. The *raid force commander* approves the results of the mission analysis. He makes the final decision on what tasks are essential and provides the restated mission. The mission statement provides the focus for subsequent staff planning.

STUDYING ALL AVAILABLE INFORMATION

Mission analysis, in part, determines the information known to the raid force commander and that which must be determined. All infor-mation is then studied to determine the course of action that promises the best chance of success. Planners initially analyze current information on the enemy situation, the objective area (terrain, weather, hydrography, local communication systems), and the attitude of civilian inhabitants. Recent, concurrent, and imminent operations by other friendly forces and the effects on the enemy are considered in detail. The requirements of stealth, speed, and surprise generate a need for a thorough intelligence preparation of the battlefield.

Sufficient information on the enemy and the area of operations is required to evaluate the adequacy of the proposed raid force as organized and equipped, and to predict its relative probability of success or failure. A force assigned a raid mission has to take calculated risks, but planning a raid based on sketchy or unreliable intelligence increases the probability of failure.

Sources and Agencies

Planners at each level of command carefully analyze available information to ensure that it is sufficient. Requests for additional information are initiated as soon as the need is determined. The request must be expressed in precise terms to minimize collection efforts. Collection plans provide for continued gathering of information on the raid objective area until the raid is executed. Verification of information and operations security are fundamental.

Intelligence Requirements

Intelligence requirements encompass terrain, weather, and hydrography; enemy defenses and patterns of action; enemy reaction capabilities; and evasion planning. Intelligence must be presented in products of consumable form such as terrain models, photographs, overlays, and easily understood reports, tables, and matrices.

Terrain, Weather, and Hydrography. Intelligence requirements relating to terrain, weather, and hydrography in the target area include--

· Photographs of landing sites taken at low and high tide showing beach approaches and exits. When practical, ground photographs, views from boats, and periscope photography are obtained.

· Detailed weather forecasts including cloud and fog coverage for the designated period and the probable effects of weather on the actions of the raid force and the enemy.

- Probable effect of expected wind, surf conditions, terrain, and vegetation on the ability of the enemy to detect the raid force.

- · Location of possible beaches and/or zones and relative evaluations of each.

- Detailed descriptions of routes to the target area and the relative advantages and disadvantages of each.

- Description of the nature of any possible obstructions or hazards on routes and beach approaches, including nature of terrain and soil, local currents as affected by the tide, types of bottom (lava, coral, or sand), and precautionary measures which may be taken to circumvent them.

- Location and description of possible hides or caches for equipment and men in the vicinity of assembly areas, landing sites, and/or LZs.

- Detailed description of local surf conditions to include any navigational hazards which may impede the ship-to-shore movement.

- Additional topographic maps and hydrographic charts or other printed information available on routes or beach approaches.

- Requirements for special equipment generated by terrain and weather in the area of operation; e.g., climbing equipment.

- Location of any soil or vegetation in the area that could prematurely reveal recent passage of the raid force; e.g., a muddy road surface, and indications of best places to cross roads to avoid detection.

Enemy Situation. Intelligence requirements relating to enemy defenses provide the basis for detailed analysis of enemy defenses in and around the target area to include--

- Strength and arms of enemy forces.

- Plans and intentions of the target installation in as much detail as possible.

- Nature and location of guard posts and routes; composition and frequency of patrols, including using sentry dogs.

- Enforcement of continuing actions.

- Emplacements, trenches, barrier systems, and fields of fire.

- Enemy communications systems.

- Existence of illuminating devices and intrusion detection devices, including mechanical, electronic, magnetic, infrared, seismic, and acoustic.

- Details of enemy routine, including off-duty routine.

- Purpose of all buildings and installations.

Enemy Reaction Capabilities. Intelligence requirements pertaining to the enemy's reaction capability demand a realistic estimate of the--

- Size and location of forces the enemy has available to reinforce at the objective.
- Speed with which the enemy can react once the raid is discovered.
- Routes which are available for this purpose.
- Course(s) of action which the enemy is most likely to adopt. When enemy drills have been observed, the report includes a description of enemy tactics and defensive plans.

Evasion Planning. The G-2/S-2 should issue evasion planning guidance on actions to be taken by personnel separated from their unit during operations ashore and the plans to attempt their recovery.

Dissemination

The raid force commander is kept abreast of the ongoing intelligence collection effort. He must closely monitor current and imminent friendly operations that may affect the execution of the raid. All aspects of the raid must be familiar to the raid force as the effects of an unfamiliar environment detract from efficiency and cooperation within the force.

Counterintelligence

Detailed counterintelligence planning and active counterintelligence measures are required in support of raids.

Strict security, secrecy, and isolation of the raid force are required during planning, training, rehearsals, and movement to the target area.

DECEPTION

The raid operation should always contain plans to deny the enemy knowledge of the raid force's insertion, or to deceive him as to the location and intent of the operation. Deception operations are planned concurrently with development of the raid order to ensure it supports the commander's concept and can be logistically sustained. Feints, false insertions, and other events all add to the raid force's deception plans. Selection of unexpected means, time, place, and routes of insertion, as well as use of speed and mobility, give less reaction time to the enemy. All events must be coordinated to ensure integration with the higher organization's deception plan.

Deception techniques that may be used include--

- Multiple helicopter and/or waterborne landings.
- Dispersion of insertion craft (air or water), both in time and space.
- False landings and radio or message traffic to disguise actual insertions.
- Fires such as air strikes and naval surface fire support in other than the objective area.
- Increased reconnaissance flights by aircraft or unmanned aerial vehicles over false targets.
- Electronic attack.

OBJECTIVE AREA ACTIONS

Objective area actions; i.e., those actions required to accomplish the purpose of the raid, are developed. Designed to maximize surprise, speed, shock, and simplicity, they provide the basis for planning of other critical events such as insertion, withdrawal, and isolation of the objective by raid force elements and fire support means.

The plan for the attack of the target establishes the planning requirements for all other critical events. The raid mission establishes the general scope of action during the attack on the target; e.g., whether total or partial destruction is to be accomplished, removal of specified equipment, capture of prisoners, number of evacuees, etc. The raid force is organized into task elements to achieve maximum flexibility and to reduce troop requirements to the essential minimum number. Each group is precisely tailored to accomplish a specific task in support of the general plan of attack. As a minimum, forces are usually organized to eliminate enemy security, attack the target, and cover the with-drawal. Plans for attack of the target are simple, easy to execute, and within the capabilities of the raid force.

Timing of all critical events (most importantly objective area actions) is situation-dependent. Raids that support other operations may require detailed timelines to ensure the desired effect for the larger operation is produced. Normally, movement and objective area actions are conducted under the cover of darkness. Raids that require exten-sive movement may necessitate the use of hides or harbor sites during daylight hours. All times should be checked and verified in a number of rehearsals approximating actual conditions expected during a raid.

WITHDRAWAL

The withdrawal normally starts immediately after mission accomplish-ment. The withdrawal is carefully planned. At this stage of the operation, the intensity of the raiding troops' offensive spirit tends to decline. The enemy is now alert and, depending on the mission, the raid force may be burdened with wounded, captured enemy, and captured equipment. For these reasons, the commander ensures that the withdrawal is executed rapidly and systematically.

Raid elements withdraw along predesignated routes and in a specified sequence except when the raid commander orders a change. Plans include provisions for certain elements to cover the withdrawal and to conduct delaying actions if the enemy pursues. Maximum use is made of available artillery, naval surface fire support, and aviation support to cover the withdrawal and, if necessary, reembarkation. Plans for the withdrawal include alternate provisions as to time, routes, and sequence of movement to friendly lines, reembarkation points, and/or landing zones. Accountability is paramount and planned for throughout this stage of the operation.

Factors considered in planning the withdrawal are the time required to attack and accomplish the mission, the most probable and most dangerous enemy reaction, and the foreseeable casualty load. Routes to the reembarkation point are selected which facilitate movement, are easily identified, and frequently change direction, thus difficult to predict. Preferably, they are behind terrain that affords protection to personnel withdrawing and offer several good positions from which security elements may block enemy counteractions. Primary and alternate withdrawal routes cannot include a mere reversal of the route to the objective. This route may have become the focus of enemy security activity.

The withdrawal plan is flexible, to include alternate provisions as to both time and place. Special situations may permit planning for the withdrawal of the raid force directly into territory of either friendly conventional or guerrilla forces. Withdrawal by air is considered when facilities and aircraft are available.

In amphibious withdrawals, surf characteristics are evaluated in the same manner as for landing the raid force. Tides and sea state should be considered with regard to their effect on landing craft, landing craft air cushions (LCACs), rigid raiding crafts (RRCs), and combat rubber raiding craft (CRRCs). Their effect on the surf are considered when selecting an exact time for withdrawal. A key factor in determining the suitability of environmental conditions is calculating the modified surf index. The *Joint Surf Manual* (CNSP/CNSL INST 384748888740.1) explains how to calculate the modified surf index and the operating parameters of landing craft.

Alternate plans for withdrawal include provisions as to routes and sequence of movement during the withdrawal, alternate methods of shore-to-ship movement, or air withdrawal. Alternate times are selected in the same manner as primary times and usually occur at approximately 12- and 24-hour intervals.

Withdrawal and reembarkation continue until the last man and piece of equipment are in friendly territory or are reembarked. The psychological impact is much more effective if the raid force disappears without a trace. However, equipment that cannot be withdrawn is destroyed.

FIRE SUPPORT

The tactical situation will dictate the fire support available to the raid force. Range, other operational demands, and collateral damage concerns may limit the fire support dedicated to the raid force. Surprise, speed, superior mobility, and improvisation can obviate some fire support requirements. However, detailed planning and integration of available support is essential. The general principles on using artillery, naval surface fire support, and air support during normal operations apply to the raid. Fire support plans must be coordinated with development of the landing plan, movement to the objective, objective area actions, withdrawal, and the plan for reembarkation.

Although fire support planning is basically the same as that for other types of attacks, it is generally more detailed in raid planning. To achieve surprise, fire support may be withheld during certain phases. Nevertheless, fire support, particularly on-call fire support, should be considered for every phase. In the event of premature detection, fire support may be the most rapid and effective method of reinforcing the raid force.

Means

All available fire support assets should be considered (see FMFM 2-7, *Fire Support in Marine Air-Ground Task Force Operations).* However, the environment and distance to the objective from friendly units may preclude the employment of some assets; e.g., an amphibious raid may not allow for the employment of artillery support. A long-range, inland raid may likewise preclude employment of mortar or naval surface fire support. Close air support (CAS) aircraft and EA-6B electronic attack aircraft, however, are capable of supporting most raid scenarios. Multiple fire support assets should be employed to ensure adequate coverage and support for the raid force. This may require displacing fire support assets, such as artillery or mortars, well-forward to range the raid force's route, its objective, and any enemy reinforcements or counterfire systems. When forced to rely solely on aircraft for scheduled or on-call fire support, raid planners should avoid any gaps in scheduled aircraft on-station times. To effectively employ available fire support means, the raid force should include forward air controllers (FAC), artillery or mortar forward observers, naval surface fire support spotters, and/or aerial observers, as appropriate.

Movement to the Objective

Because the success of a raid usually requires surprise, raid forces will generally not employ any preparation fires on the objective or other fires in support of their movement that might alert the enemy. However, pre-H-hour fires may be an effective part of deception operations to draw enemy attention away from the raid force or its objective. On-call fire support should also be planned to support the raid force if it is detected en route and requires assistance to break contact, conduct an emergency withdrawal, or continue to the objective.

Assault on the Objective

Fire support may be scheduled to isolate the objective from enemy reinforcements or to attack known enemy indirect fire weapons capable of striking the raid force. Raid planners should consider dedicating on-call fire support to provide responsive counterfire against enemy indirect fire weapons that engage the raid force and to attack any lucrative targets that may be exposed during the raid.

Withdrawal

Fire support may be required (indeed may be critical) to withdrawal and recovery. At a minimum, raid planners should consider dedicating fire support assets for preplanned, on-call missions to assist the raid force in breaking contact and to prevent interference with its withdrawal.

Coordination

For most small unit raid forces, the raid force commander generally serves as his own fire support coordinator. However, fire support coordination centers of higher headquarters and/or the supporting arms coordination center will be closely involved in raid planning and execution and should be prepared to assist in providing and coordinating on-call fire support assets and fires. Standard fire support coordination measures should be employed in raid operations, although these measures may be positioned very close to friendly forces (see NWP 22-2/FMFM 1-7, *Supporting Arms in Amphibious Operations,* and FMFM 6-18, *Techniques and Procedures for Fire Support Coordination).*

Raid planners should consider employing and positioning fire support coordination measures to preclude any uncoordinated fires which might alert the enemy to the raid force or cause him to change his defensive posture. This can also be accomplished by establishing rules of engagement which restrict attacks by fire support units or aircraft prior to assault. Fire support coordination measures can also enable fire support units and aircraft to rapidly engage targets of opportunity that could interfere with the withdrawal of the raid force.

TASK ORGANIZATION AND EQUIPMENT LIST

After detailed plans are developed, requirements for troop units, specialists, and support can be accurately determined. Raids are normally of short duration. Unless operating in extreme climes, the raid force may be lightly equipped and require minimum logistics support. The raid force is organized into specific elements/units/teams commensurate with the number and nature of tasks to be accomp-lished. Tables of organization are adhered to as much as possible. However, some deviations are inevitable because elements are tailored to specific tasks assigned. Based on the organization of the raid force, lists of specific equipment, weapons, ammunition, and supplies to be carried by each subordinate group are prepared.

CHOICE OF LANDING PLACE OR DEPARTURE POINT

Beach Selection

All beaches are examined to support objective area actions and determine the best approach to the target. Calm conditions are desirable although movement through these waters may increase the difficulty of effecting surprise. Sheltered waters are usually enclosed in a bay or in the lee of islands. Ships and other surface craft run a risk of detection as they approach and interception as they return. Sheltered waters are also more prone to minings. The initial point for raid force orientation ashore must be determined. The location must be known precisely with respect to the beach/zone and to the objective. The landing beach/zone itself should be large enough to allow for errors in predicted drift of swimmers, boat teams, or air crews. Normally, landing places are avoided which cannot be approached from several different directions. Other factors considered in beach selection include surprise; enemy dispositions; distance to target; sea approaches, beach characteristics, and beach exits; and equipment to be moved over the beach.

Physical Characteristics

Selecting a landing beach usually involves a compromise between a beach which permits easy landing, with some risk of detection, and one which provides maximum security at the risk of a difficult or hazardous landing. Examples of each are the wide, flat, sandy beach backed by an easily traversed hinterland and the narrow, steep, rocky beach backed by cliffs.

A predominant factor to be considered is surf state. Hydrographic offices should be consulted for up-to-date information. Hydrographic surveys and/or confirmatory beach reports may be required before the raid.

Surf characteristics on a given day are based on the nature of the bottom, direction and speed of the wind, the distance between successive swells (wave length), the state of the tide, and the nature of currents. Offshore shoals, ledges, and rough bottom contours tend to reduce the surf. Offshore islands tend to break up ocean swells and produce several patterns of smaller waves. Kelp or dense seaweed reduce wave height. Swift currents flowing in the direction of wave advance and onshore winds reduce wave height.

A reef face or other abrupt break in the bottom may cause each wave to break up into smaller waves. A submerged ridge perpendicular to the coast increases wave height; conversely, a submarine canyon reduces wave height. A steep bottom causes waves to break rapidly and close to or directly onto the beach accompanied by violent wave reaction. A flat bottom causes waves to break gradually and at a greater distance from the beach with several foam lines being formed between the breaker line and the beach.

A sand bar parallel to the beach causes waves to peak up or break depending on the depth of water over the bar. A single breaker line may form over a bar while another breaker line forms closer to or on the beach. Several bars may cause multiple breaker lines. Sand bars are frequently found off sandy beaches exposed to wave action.

Landing Zone or Drop Zone Selection

The raid LZ or drop zone must support the planned actions at the objective.

The raid force can land on or near the objective and seize it before the enemy can react. This avoids forced marches over land carrying heavy combat loads. If there is no suitable landing area near the objective or the enemy has a strong reaction force nearby, this option is not favored.

The raid force can land unseen far from the objective. It then assembles, reorganizes, and moves into an objective rally point near the objective. The objective is seized after security and support elements are in place. This option may ease coordination by allowing a more complete orientation of the force before engaging the enemy.

EMBARKATION AND MOVEMENT TO THE OBJECTIVE AREA

Embarkation and movement to the objective area are planned in accordance with standard amphibious procedures. Speed and/or movement in periods of reduced visibility are emphasized to maintain the element of surprise. Appropriate antiair warfare and antisubmarine warfare measures are planned to protect the attack group.

SHIP-TO-SHORE MOVEMENT OR DEPARTURE LANDING ZONE TO LANDING ZONE

Planning for the ship-to-shore movement is similar to that for the amphibious assault. The CATF and CLF maintain positive control of ship-to-shore movement through control groups and temporary organizations. Execution of the waterborne and helicopterborne ship-to-shore movements is delegated to subordinate commanders.

The method selected for the ship-to-shore movement should be that which will land the raid force with the least probability of detection, as close to its objective as possible, and as simply and rapidly as possible. This is but one more mitigating factor in favor of smaller raid forces. Landings should take advantage of night and conditions of reduced visibility and detection. The raid force emphasizes silent landing techniques which preserve surprise. Care must be exercised in the introduction of advance reconnaissance elements designed to facilitate landing of the raid force and/or conduct target surveillance. Capabilities and benefits provided by employment of these elements must be weighed against the increased risk of compromise.

Primary control agencies are the Navy control group and tactical air control group, both of which may be tasked with control and/or supervision of the ship-to-shore movement. However, due to the simplicity of plans, the smaller scale of the raid, and proficiency gained during rehearsals, control ships, lines of departure, and other control measures are seldom used. Ship-to-shore movement is rehearsed in detail under conditions similar to those in the objective area to ensure proper timing and coordination. The same principles apply to movement by air or surface means.

Detailed guidance for the conduct of waterborne and

helicopterborne ship-to-shore movement is found in chapters 4 and 5 in NWP 22-3/FMFM 1-8. This manual describes the larger amphibious operation. Descriptions of fundamentals and considerations generally apply to raids. Control groups and temporary organizations may apply depending on delivery means, raid force size, distance to the objective, and other METT-T factors.

MOVEMENT FROM THE ASSEMBLY AREA, BEACH, OR LANDING ZONE TO THE OBJECTIVE

Plans are made to ensure that the assault element can reach the target intact. The security element is employed to neutralize and contain any enemy opposition that attempts to interfere with the assault element. Composition of the security element depends on the task assigned and the expected resistance it will encounter.

REEMBARKATION

A reembarkation point or points are selected during planning. While a raid force may reembark at the same point it was launched, another location may be required to enhance security. Variation in transportation means (necessary when enemy air or naval action warrants change) may require alternate reembarkation or pick-up points. Pro- visions are included in raid plans for their use in emergencies. Accountability and marshalling of all hands is especially important and planned for. Specific escape and evasion plans are made for subsequent recovery of persons unable to reembark with the main body of the raid force.

SELECTION OF DATE(S)

Tidal ranges vary widely based on geographic location and moon phase. It is most desirable to employ the raid force at the first combination of suitable moon and tide. Normally, this will provide 3-day employment windows of reasonably stable conditions. When the force depends on waterborne transportation through reembarkation, this 3-day window will be considered for insert, extract, and an alternate day for bad weather.

Generally, it is better to land with a rising tide to reduce the stranding of landing craft. However, defensive arrangements are usually designed to defeat landing parties somewhere near the high watermark. Landing at low tide may provide the raid force a measure of surprise by maintaining some distance from enemy local security posts and avoiding underwater obstacles designed to impede watercraft. The chief disadvantage of a landing at low tide is the loss of time. It may add as much as a mile of open terrain to movement requirements.

The moon helps most when it is due to rise on the landward side soon after the raid force is ashore. This allows ships and boats to approach in complete darkness with landmarks faintly silhouetted against the glow which precedes the rising moon. Illumination and moon angle will also impact on aviation operations and, in particular, must be considered for helicopterborne insertions and extractions.

Special attention must be given to seasonal and climatic conditions of the geographic location. Concealment can be greatly altered by seasonal changes, which impact the ability of the raid force to maneuver. Weather has a direct effect on helicopter support to the raid force and on the ability of the raid force to move through the terrain.

Additional geo-political factors may impact on determining the date for a raid. Considerations center on whether to choose a date which will maximize international political exploitation or that which will minimize the coverage to local dissemination; e.g., the timing of national and local elections. Local traditions and festivities; i.e., national or local holidays, and social mores and customs of the target area; e.g., religious holidays or work schedules are also considered.

TIME-SENSITIVE PLANNING

Operations may have to be conducted on short notice or *immediately.* These raids require a rapid planning cycle to expedite execution. Rapid planning incorporates the same procedures, albeit in a condensed timeframe, shown in appendix A. Rapid (or time-sensitive planning) does not ignore the requirement for detailed planning. However, it relies on established procedures, a well-trained force, and a focused effort. The technique of time-sensitive (rapid/compressed) planning has been developed with the aim of allowing the raid force commander to commence execution within 6 hours from receipt of the mission.

Key elements of rapid planning include--

- Reverse planning.

- Each commander uses the 1/3 - 2/3 rule, allocating one-third of the available planning time to himself and two-thirds to immediate subordinates.

- Critical, accurate intelligence. Priority information requirements and critical go/no-go questions should be formed as soon as possible.

- Fire support planning.

- Maximum use of checklists, SOPs, and specific force modules. Eliminate forces not required.

- "Play books" developed from prior detailed planning. These cover most likely or typical missions, and require only slight modification to meet the requirements of a specific situation.

- The force commander's confirmation briefing. This is extremely important. Coordination problems must be resolved on the spot. Brevity is essential.

(reverse blank)

Chapter 4
Logistics

The raid force commander must consider his support requirements before, during, and after the operation. He must evaluate his mission, the situation, his task organization, equipment, and weapons necessary to accomplish his mission. During preparation, the force may require special munitions, personnel with unique skills, and transportation. Unlike that support provided to extended operations, combat service support of raid operations is short-term, front-end loaded, and narrow in focus. Combat service support planners and logistics support facilities must have specific guidance and detailed prioritization.

SUPPLY

The raid force commander determines the forces's requirements for all classes of supply, storage, salvage, and disposal necessary to support preparations for the operation and that which must be in place during the operation. When time is available for extensive rehearsals and other preparations, supply requirements to support the force before the operation may be significantly greater than at any other time. For example, a raid force will rarely require class IV items during the operation, but may require extensive construction materials in developing rehearsal sites and target mockups. Supplies carried by the raid force are restricted to that established in planning and tested in actual rehearsals. Ordinarily, the raid force is not resupplied during the operation. Emergency supplies, primarily class V, may be necessary if the enemy disrupts the withdrawal of the raid force.

MAINTENANCE

All items of equipment must be in satisfactory repair. Weapons and equipment must be inspected, adjusted, modified, or replaced as necessary before and after each rehearsal and before execution. Maintenance personnel with requisite spare parts and replacement items will rarely accompany the raid force during execution, but may be assigned during preparation. Standard equipment may be modified as necessary to cope with the special conditions of its use during the raid. Special equipment may be improvised or manufactured when necessary to meet special requirements generated by the nature of the raid. The raid force *must* have the exact types of equipment required to carry out its tasks ashore.

TRANSPORTATION

The raid force may be transported into and out of the objective area by naval, air, and/or ground transportation. The effects of terrain, weather, and enemy capabilities require detailed coordination of transportation support to ensure the unhindered movement of the force. Rehearsals should be conducted with the actual transportation means assigned for the operation. Spare vehicles/aircraft should be allocated to account for attrition during training and execution of the raid.

Transports which can approach at relatively high speed and load/unload rapidly are used for large-scale raids. Small units may use transport aircraft, helicopters, landing craft, small boats, and parachute. Lightweight, air-transportable vehicles may enhance mobility and permit the raid force to move rapidly into position, execute its attack, and withdraw from the objective. When insufficient vehicles are available to move the entire raid force, the raid force commander may permit their use by selected teams or force elements to capitalize on this mobility. The nature of most raids limits the requirement for ground transportation at the objective.

HEALTH SERVICE

Health services ensure each Marine is physically capable of accom-plishing the mission. The level of medical support available before execution is the same as any other unit. Preventive medicine is significant during preparation to minimize last minute replacement requirements and prevent the effects of regional endemic diseases. During the operation, medical care may not be any other than that provided by the unit hospital corpsman. Evacuation plans must be established for wounded or others that may impede the ability of the force to execute the mission or conduct the withdrawal. Casualties are evacuated as early as practical by the fastest means available.

(reverse blank)

Chapter 5
Training

Only the well-trained raid force can be expected to meet the demands of raid operations. Landing force, Navy, and aviation elements detailed for the raid operation conduct integrated training as early as possible. One of the main objectives of training is to rehearse the raid under conditions approximating, as nearly as possible, those which will be met during the operation. Every effort is made to rehearse the raid on coastline and terrain most comparable to the actual objective area. The aim is to test the plan, train the force, and familiarize them with the area over which they are to operate.

INDIVIDUAL TRAINING

Individuals are selected for the raid force based on their experience and special skills required for accomplishment of the mission and/or specific tasks. Before assignment, personnel are screened for physical or medical problems which would hinder their performance. Individuals must be physically able to accomplish a mission calling for the utmost in physical stamina. When possible, the raid force begins training slightly overstrength to allow for attrition during training. Individuals separated from the raid force during training as well as administrative support personnel are kept in isolation until the mission is completed.

Separation of individual training and unit training is minimized. Individual skills are improved while undergoing unit training. Refresher training in basic subjects, especially applicable to the raid, is often necessary. This may include amphibious training; swimming; helicopterborne operations; hand-to-hand combat; evasion and escape (E&E) training; and survival training. Special training is provided in small boat handling, demolitions, use of enemy weapons and equipment, cliff-scaling techniques, and operation of special equipment and arms, as required.

Subjects in which special training is most often necessary include night operations, obstacle recognition and breaching, physical training that develops stamina, use of communication means to be employed during the raid, close quarter battle, and intelligence collection. Each individual must be thoroughly familiar with his own role and the functions of his fellow Marines in the operation and how they relate to the over-all plan. Cross-training of all personnel is maximized. It is critical to reducing the size of the raid force, the ability to execute contingency plans, and to the success of the raid force if casualties are sustained. The difficulty of executing critical events during the raid (in particular a withdrawal at night) can be mitigated through individual training.

UNIT TRAINING

Unit training is the most important ingredient to successful team-building. Unit training includes both element and force training. Rigorous and thorough training develops special skills, unit cohesiveness, and esprit required to conduct successful raids. SOPs are established and tested. Emphasis must begin with actions in the objective area and be followed by all critical events to ensure successful completion of the mission and recovery of the force. Elements conduct internal rehearsals of assigned tasks. The use of special equipment must be practiced and perfected. Demanding unit and individual standards must be developed and adhered to.

Staff training must not be neglected. Command post exercises, map exercises, and raid force rehearsals all provide opportunities to prepare the raid staff. Training must be continuous for all personnel from pre-mission training, briefings, and rehearsals to mission execution. Generally, units and commanders are selected for their experience as a coordinated team, their morale, and their previous combat experience.

Integrated training with specialists, Navy units, and supporting forces commences as early as possible. Additional training may be required to adjust the raid force to special climatic conditions of an objective area when it differs materially from the one in which they normally operate.

During the training cycle while the individual is improving his skills, the unit specifically trains for its mission. Troops are uniformed and equipped in the exact fashion required for the raid. The raid force should train against aggressors to refine their skills against an enemy and to test their individual proficiency. However, the aggressor unit may require quarantine to ensure operations security. Training frequently extends to periods where raiding troops are subjected to extreme physical and mental strain.

When time permits, elements are cross-trained in the functions of the other elements.

REHEARSALS

Thorough, integrated rehearsals are required to obtain precision and speed in the execution of a raid. All participating forces are drilled in every detail of debarkation, movement ashore, operations ashore, withdrawal, and reembarkation. Timing and synchronization of effort cannot be accurately estimated without adequate rehearsals of the entire raid force. The purpose and conduct of rehearsals for raids are similar to that for the assault. They are conducted to verify time and space factors for each phase of the raid; to familiarize personnel with the plan and special equipment and techniques to be employed; to measure proficiency of the individuals, elements, unit, and group; to develop coordination, resolve problems, and suggest necessary changes; and to incorporate all previous modifications to the plan.

Requirements

The success or failure of the whole operation depends on the quality of rehearsals conducted before the operation, and every conceivable effort to achieve realism enhances the probability of success. Drills progress in scope from terrain model walk-throughs to full-speed live fire rehearsals. A full-scale replica of the target area is constructed, roads and trails taped out, and buildings represented. Rehearsals are conducted under the conditions expected; e.g., at night or under reduced visibility, from over-the-horizon, and against an active enemy. Movements from point to point are carefully timed. Anticipated contingencies and simulated casualties to key personnel must be covered.

During rehearsals, requirements for special equipment, such as scaling apparatus and/or special demolition kits, are normally generated.

Testing of Plans

During rehearsals, the plan for the raid is validated and refined. The plan of attack is examined to ensure--

- The plan for actions in the objective area provides for efficient and rapid mission accomplishment without delay or confusion.

- All objectives contribute to mission accomplishment.

- The plan is feasible under conditions expected.

- The plan maximizes surprise.

- The plan is flexible so that with a loss of surprise, a smooth transition can be made from a plan based on stealth to one based on firepower.

- H-hour and/or L-hour are valid.
- The raid force is the proper size.
- Existing tables of organizations are used as much as possible.
- All available resources have been exploited.
- Tasks assigned to subordinate elements are within their capabilities.
- An adequate reserve is provided for execution of contingency plans and to take advantage of unforeseen opportunities.
- The most dangerous and most likely enemy reactions have been considered.
- Selected landing areas best support accomplishment of the mission.

- Time is available to land, attack the target, and withdraw without subjecting the raid force to unnecessary risks.

- The plan for withdrawal allows the force to quickly break contact and move undetected even if initiated under pressure.

- All essential elements of information (EEIs) and other intelligence requirements (OIRs) are identified.

Critiques and After-Action Reviews

Pre-mission planning and rehearsals should include the use of critiques and after-action reviews. It is unreasonable to expect the observations of *only* key personnel to provide the necessary basis for plan validation. After-action reviews (conducted with all hands after each training milestone or rehearsal) will provide the detailed feedback the raid force commander requires. Every leader and individual must understand his mission thoroughly and be required to brief the next higher leader on the exact scope and purpose of the assigned mission and how it fits into the overall plan. This action ensures that each Marine understands the mission, both from his own perspective, as well as the overall perspective.

Appendix A
Time-Sensitive Planning for the Raid Force Commander

Rapid planning allows the commander to develop and disseminate decisions under limited time constraints. This appendix highlights common raid force concerns. Procedures are valid for all raids and can be used with appendixes C through G for specific raids. Rapid planning techniques are emphasized when a force is assigned a mission requiring an immediate employment of forces. Six hours after receipt of the warning order, Marine units must be capable of executing the operation. Six hours is the standard. The force commander must complete a detailed, coordinated plan and then continue to refine that plan before execution. This is a difficult task and impossible if the commander and staff do not have a high degree of situational awareness.

SITUATION DEVELOPMENT

Situation development is a continuous process. During peacetime, operational organizations and supporting agencies must keep abreast of world-wide events. Anticipation of potential missions and the wargaming of each provides the staff the momentum to quickly develop specific plans upon receipt of a warning, alert, or execution order. Without "leaning forward," the commander and staff must plan from a cold start. To do so and still meet the 6-hour standard requires a significant increase in risk--additional risk that is unnecessary.

TIME ALLOCATION

The argument for the capability to conduct rapid planning is based on time. Time cannot be saved or stored. Just as we will not waste ammunition, fuel, or water, we will not waste time. The effort in thought cannot be less than that put forth when more time is available.

The key is to quickly recognize the problem, develop a viable solution, and refine that solution to reduce risk to an acceptable level. Rapid planning is part of rapid preparation.

Rapid preparation is the total package required to execute a mission within the 6-hour standard. It includes planning, rehearsal, and logistic preparations. The force cannot wait for the completed plan before conducting preparations. SOPs, operational checklists, and initial guidance allow preparations to start concurrently with plan develop-ment. The faster the raid force commander and his staff can produce a sound, viable operation plan, the quicker supporting plans and preparations will be integrated with it.

A well-known technique is the "1/2 rule." Better still is the "1/3 - 2/3 rule." The key point is to discipline or otherwise limit yourself to half or one-third of the available planning time, allocating the remainder to your immediate subordinates. For example:

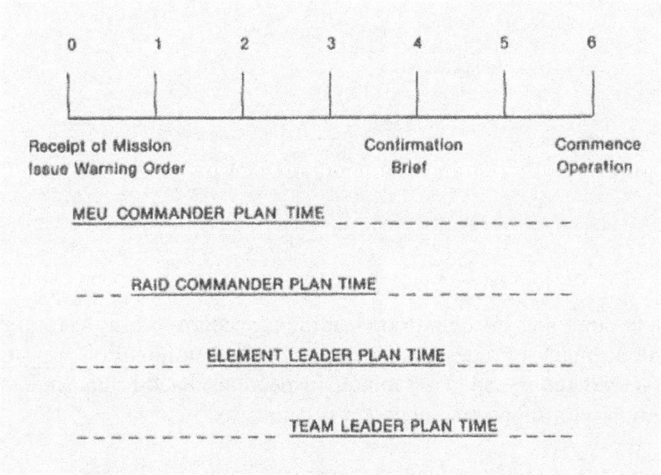

Likewise, subordinate raid force elements must plan the use of available time to ensure mission readiness at the prescribed time. In the time-sensitive environment, notional or standard procedures described in SOPs serve as useful guides to rapid preparation. Preparation by the ACE for a helicopterborne raid may illustrate this requirement. Detailed planning by the ACE, including fire support planning; preparation of kneeboard packets; forward arming and refueling point(s) when necessary; and tactical recovery of aircraft and personnel plans is concurrent with these activities. For example:

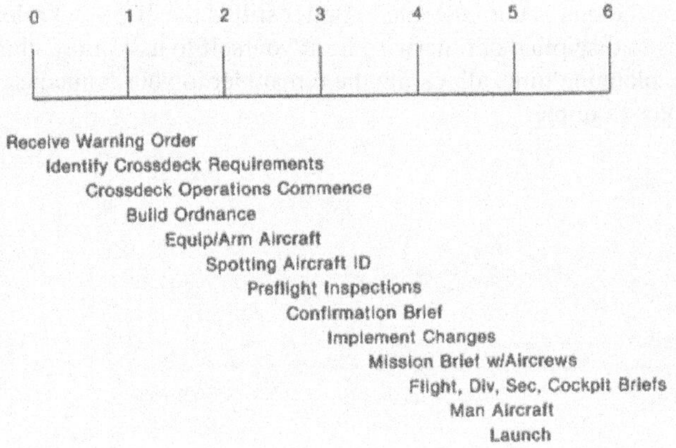

The commander continues his own troop leading procedures during that time allocated to subordinates. Predesignated equipment and ammunition packages are staged and issued. Last minute inspections, back-briefs, and final coordination with supporting agencies is conducted.

One of the commander's most critical concerns is accurate intelligence. Predominately, intelligence collection in support of raid planning will be conducted by agencies outside the raid force. The commander and his staff can focus this collection effort by identification of EEI and the commander's priority intelligence requirements (PIRs). The prudent commander will articulate to his staff those items of information he considers critical to his decisionmaking process for a type-operation. When tasked, this standing list of information requirements and priorities may require some refinement, however, the baseline has been established.

Go/no-go criteria is determined as soon as possible. These are critical conditions that must exist to conduct the operation. Each stated criteria reflects a risk the commander will not accept. They may be related to intelligence, material readiness, political conditions, or training status at any point in the operation. Should any go/no-go criteria not be met, the raid is delayed or cancelled. For this reason, these criteria must be thoroughly wargamed and approved by the tasking headquarters.

FORCES AVAILABLE

Organizations should develop specific force packages for each type of raid. Upon receipt of a warning order, the basic raid force, reflected in a force package, is readily identified. This force is then modified as required by that specific mission. As a matter of routine, the forces designated in each package must have conducted the integrated training required before execution. When forward deployed afloat, forces within each package should be embarked together.

OTHER TOOLS

Commanders are limited only by imagination in developing methods and procedures for reducing preparation and planning time. However, these procedures must be thoroughly familiar to the staff and any other planners. The key is training. Vehicles for training include map exercises, field situational training exercises, tactical exercises without troops, and force training exercises.

Execution checklists are a must for time-sensitive planning. They are event-driven and are used to display critical events and decision points to the entire MAGTF. They may be used to assist in reporting both up and down the chain of command. Codewords for selected events may be provided. Appendix B provides a sample execution checklist.

Standardized procedures set forth in unit SOPs provide a familiar methodology for planning and execution. These procedures are under constant scrutiny to ensure applicability to threat, technology, and force structure. Experience has shown that simple procedures thoroughly understood by all are of greater benefit than large, complicated SOPs that attempt to cover every detail of an operation. Some examples of SOPs that may be developed include helicopterborne unit and heliteam procedures, reconnaissance and surveillance procedures, bump plans, and inspection procedures.

SEQUENCE

The planning sequence applied by the commander and staff in the development of the operation order must be reviewed to ensure appropriate regard is given to critical events. Many tasks may be addressed in unit SOPs and will only require a brief analysis. Well-trained units will dedicate the majority of the available planning time to mission-unique events. These units have developed a systematic approach to problem identification and solution. One approach is to determine the answers to a variety of pertinent questions arranged in a sequence that leads to construction of the plan. An example of this approach is provided below.

Examine all available information regarding the enemy and the area of operations. The study of the area of operations should provide critical information dealing with alternate routes and E&E procedures. Attempt to determine--

- Organization, equipment, composition, and disposition of enemy forces in the objective area.

- Enemy reaction capability including type, size and location of reaction force; reaction time; avenues of approach; and most probable and most dangerous courses of enemy action.

- Intelligence requirements. Identify critical unknowns and express them as EEIs or OIRs.

Plan the action on the objective. Determine--

- Security procedures.
- Movement from objective rally point to assault positions.
- Assault position locations and control measures.
- Signals and communications.
- Fire support and electronic warfare.
- Special team tasks and sequence of actions.
- Alternate routes, control measures, and fire support plans to support contingency plans.

Plan the movement from the launch platform to the objective area. Plan for--

- Movement to and organization at the objective rally point.
- Primary and alternate routes.
- Supporting fires.
- Implementation of the bump plan.
- E&E procedures.
- Clandestine ship-to-shore movement.
- No-comm plan.

Plan the withdrawal. Include--

- Primary and alternate routes.
- Link up with security/supporting forces.
- Maximum use of supporting fires.
- E&E procedures.
- Casualty handling and evacuation procedures.
- Enemy prisoner of war handling procedures.
- Accountability procedures.
- Recognition signals.
- Alternate means of extraction.

Prepare the force size, composition, and task organization. Keep the force as small as possible and delete unnecessary equipment.

COMMANDER'S CONFIRMATION BRIEFING

The commander and staff operating in a time-sensitive environment do not have time to conduct the number of briefs that normally take place during deliberate planning. Upon mission receipt, the commander must articulate his intent and planning guidance. Guidance should include a specific planning timeline that includes the briefing schedule.

Attendees

To preclude redundant briefs, all leaders should be present for the commander's confirmation brief. The MAGTF commander will restate his intent, receive an integrated brief from all planners, and issue any last minute guidance. The commander's confirmation brief is *not* a decision brief. The raid force commander will brief his plan during the confirmation brief. Coordination problems that would not ordinarily be discovered surface during this brief and are solved *at this brief.* Once the commander's confirmation brief concludes, the plan is set in motion. No changes are made to the plan unless intelligence dictates that a change is necessary. Changes are disseminated to the entire raid force. Attendees present for a typical Marine expeditionary unit/amphibious ready group (MEU/ARG) confirmation brief include--

- The CATF, N-2, N-3, N-5, N-6, special warfare officer, and supporting arms coordinator.

- Ship commanding officers, operations officers, air boss, flight deck officer, and combat cargo officer.

- The MAGTF commander, MAGTF executive officer, S-1, S-2, S-3, S-4, special operations officer, air officer, fire support coordinator, communications-electronics officer, force reconnaissance detachment officer-in-charge, and the radio battalion detachment officer-in-charge.

- The battalion landing team commander, S-2, S-3, air officer, company and appropriate detachment commanders (engineer platoon, reconnaissance platoon, assault amphibious vehicle platoon, etc.).

- The ACE commander, S-2, S-3, mission planners, air mission commander/division/section leaders.

- The combat service support element commander and S-3.

- The raid force commander (if not one of the above).

Recommended Sequence

The confirmation brief sequence may reflect the desires of the individual commander. The following recommended sequence follows the operation order format:

- The MAGTF commander or G-3/S-3 restates the mission (specified, implied, and essential tasks).

- The MAGTF commander provides commander's intent.

- The weather rep forecasts weather and highlights all relevant factors for the objective area and sea operating areas.

- The MAGTF G-2/S-2 provides the latest intelligence update and illumination as it affects the operation.

- The raid force commander briefs specific force size, composition, and task organization; concept of operations; execution checklist; contingency and alternate plans; element responsibilities; and go/no-go criteria.

- The air mission commander briefs the aviation support plan, the launch/recovery plan, etc.

- The fire support coordinator briefs the fire support plan and organization for combat.

- The communications-electronics officer briefs communications-electronics operating instructions and critical nodes.

- Appropriate ship operations officers or combat cargo officers brief pertinent ship operations, event time lines, and the embarkation/debarkation plan.

Specific information provided in the confirmation brief is provided below. Additional information may be required for a particular mission. Likewise, some information may be omitted if unnecessary.

Confirmation Brief Format

MAGTF Commander/G-3/S-3

Specified, implied, and essential tasks

Mission statement

MAGTF Commander

Commander's intent

Weather Representative

Sea forecast

Objective area forecast

Tidal conditions

Sea state

Ceiling/visibility/temperature/winds

MAGTF S-2

Hydrography

Topography

Illumination

Civilian population

Political structure and climate

Objective area orientation (LZs/beaches/locations)

Enemy situation (state how current intelligence is):

- Strength
- Patterns of activity and routine
- Locations
- Types of organizations (as specific as possible)

- Security
- Most dangerous enemy course of action
- Most likely enemy course of action
- Air threat
- Surface-to-air threat
- E&E plan

Raid Force Commander

Friendly forces:

- Advance forces
- Rear
- Adjacent

Size, composition, and task organization

Mission

Go/no-go criteria

Concept of operations:

- Phases
- Objectives
- Subordinate unit missions
- Subordinate unit actions

Alternate plans

Insertion:

- Means
- H-Hour
- L-Hour
- Bump plan

Coordinating instructions

Accountability and marshalling

Withdrawal plan

Execution checklist

Element Leaders

Mission

Concept of operations

Bump plan

Coordinating instructions

Fire Support Coordinator

Organization for combat/assets available:

- Air, naval surface fire support
- Artillery, mortars, electronic attack, and special systems

Fire support coordination measures

Priority of fires

Scheduled fires

On-call fires

Suppression of enemy air defenses program and procedures

Smoke/immediate suppression programs

Antiair defense

Air Mission Commander

Assets/ordnance/sorties available

Event numbers and call signs

Frequencies and emission control procedures

Control points and routes

Response times and time-on-station

Troop and equipment lift:

- Helicopter employment and assault landing table/helicopterborne waves and assignment table
- Formation
- Helicopter LZ/rendezvous point/approach and retirement lanes/initial terminal guidance
- Insertion sequence and orientation
- Call signs and frequencies

- Contingency plans

Bump plan

Tactical recovery of aircraft and personnel plan

Search and rescue

Medical evacuation

Escort:

- Assets
- Call signs/event numbers
- Planned targets
- Tactics and control measures
- Time-on-station

Forward arming and refueling point(s)

Communications-Electronics Officer

Signals and codewords

Changes

Alternate communications plan

Primary and alternate nets

Frequency deconfliction

Emergency destruction plan

Appropriate Ship Operations Officers/Air Bosses

Location/point of intended movement

Operations time line

Cross deck status

Debarkation

Reembarkation

Aircraft spottings

Refueling operations

No-comm plan

MAGTF S-4 (or designated combat service support element representative)

Personnel

Supply

Maintenance

Transportation

Medical and dental support

Landing support

Data processing

Special ordnance

Raid Force Commander

Command element locations during each phase

Chain of command

Questions

Time synchronization

(reverse blank)

Appendix B
Sample Execution Checklist

Scenario: a MEU/ARG afloat is tasked to conduct a destruction raid. Plan of attack includes small boats insertion of the force reconnaissance detachment in advance of the raid force to conduct target surveillance. The raid force conducts a helicopterborne insertion, destroys the target, and withdraws with the force reconnaissance detachment for helicopterborne extraction.

EVENT NO	SITUATION	RPT	NET	FROM	TO	CODEWORD	TIME	REMARKS
1	BOAT LAUNCH	M	2	RECON	LFOC	FORD	D-1	1700
2	ABORT	X	1/2/3			ESDEL		
3	CONTINUE	X	1/2/3			INDY		
4	ENEMY ATTACK	X	2	RECON	LFOC	CHEVY		
5	BOATS CP1	M	2	RECON	LFOC	BUICK	SPLASH+30	EST
6	BOATS CP2	M	2	RECON	LFOC	JEEP	SPLASH+1 HR	EST
7	RECON FEET DRY	M	2	RECON	LFOC	TBIRD	SPLASH+1:45	EST
8	BOATS CP3	M	2	RECON-B	LFOC	GEO	SPLASH+2:15	EST
9	BOATS CP4	M	2	RECON-B	LFOC	EAGLE	SPLASH+2:45	EST
10	RECOVER BOATS	M	2	RECON-B	LFOC	TONKA	SPLASH+3:15	EST
11	SPOT REPORT	X	1	RECON	LFOC	JUDY	ASAP	
12	SALUTE REPORT	X	1/ 2	RECON	LFOC	ALICE	D-DAY 0300	
13	PLAN B	X	1/ 2	RECON	LFOC	CINDY		
14	L-HOUR CHANGE	X	1/2/3	LFCO		RITA		(+X-)
15	ALTERNATE LZ	X	1/2/1	RECON		GINA		
16	INSERT COMPLETED	M	1/2/3	RFCOMDR		KAY	L+15MIN	EST
17	ORP	M	1	RFCOMDR	LFOC	JEAN		
18	MISSION ACCOMP	M	1	RFCOMDR	LFOC	BARB		
19	ALT WITHDRAWAL	X	1/ 2	RFCOMDR	LFOC	CHRIS		
20	EMER MEDEVAC	X	1/ 2	RFCOMDR	LFOC	JOE		
21	EXECUTE E&E	X	1/2/3			TOM		
22	EXTRACT	M	3/ 4	RFCOMDR	AMC	BILL		TIME
23	FEETWET	M	2	AMC	HDC	PEPSI	RELAY TO LFOC	
24	SWITCH FREQ	X				PUNT		NET#
25	FORCE RECOVERED	M	2	RFCOMDR	LFOC	PASS		

NETS: 1 – LF COMMAND 1 (SATCOMM)
2 – LF COMMAND 2 (HF)
3 – RAID FORCE COMMAND (VHF)
4 – SQUADRON COMMON (UHF)

M - MANDATORY
X – AS REQUIRED

(reverse blank)

Appendix C
Considerations for Amphibious Raids

Marines must be able to conduct an amphibious raid on short notice at night under emission control (EMCON) conditions via helicopter and/or surface means from extended ranges. These raids are conducted to inflict loss or damage on opposing forces; create diversions; and to capture or evacuate individuals and material by swift incursion into or temporary occupancy of an objective followed by expeditious withdrawal. This is the operational standard for execution of the amphibious raid.

PRINCIPLES

Plans must be as simple as possible. Top-down planning must be conducted. Plans must create the best conditions under which surprise may be achieved. Without surprise, the raid will fail.

Training/rehearsal is mandatory. Commanders and staff must have accurate, detailed, and timely intelligence. Communications coordination is a necessity.

The raid force must be as small as possible. Do not send 100 Marines to conduct a raid that 15 Marines can accomplish.

Always plan fires in support of the raid from debarkation to recovery aboard ship. Fire planning does not require the force to execute unnecessary fires.

SEQUENCE

Gather Available Intelligence

Develop EEIs and OIRs. Consider the effects of weather, astronomical and tidal data, surf, wind and light conditions.

Use Naval Intelligence Processing System studies and maps and aerial imagery of the objective area. Analyze enemy defenses in and around the objective area.

Determine the enemy's detection capability (radar, signals intelligence, visual, and range).

Prepare detailed description of routes to the objective area.

Plan for signals intelligence, electronic warfare, and communications security.

Determine and evaluate enemy patterns of activity around the objective area. Consider enemy reaction capabilities and time. Check on any existing E&E plans.

Early Issuance of Warning Order

Determine forces to be used; staging requirements for ammunition and supplies; and planning, briefing, and rehearsal timeline.

Commander's Guidance

Guidance must be detailed and focus the staff's effort.

Courses of Action/Estimates of Supportability

Courses of action must be viable and reflect different means to success. The number of courses of action developed by the staff are limited by available time and equipment. Each staff section provides estimates.

Develop estimates concurrently with courses of action to reduce time requirements. This requires close coordination and participation between the operations section and the remaining functional area staff sections.

Concept of Operations

The plan should be simple and flexible. The concept must be within the raid force's capability. The concept must maximize surprise and stealth, and take advantage of enemy weaknesses and friendly strengths.

Embarkation Plan

Limit cross-deck requirements.

Ship-to-Shore Movement

Plan for the use of assets to get to and from the objective area in one lift. Consider alternate lift means and communications restrictions.

Consider the impact of night and reduced visibility (night vision goggles requirements and availability, deck crew requirements, and below deck lighting).

Closely monitor sea state/surf conditions. The modified surf index, found in the *Joint Surf Manual* provides guidance on launch criteria based on combat and troop loads. For surface raids using the assault amphibious vehicle (AAV), RRC, and CRRC, the modified surf index should be consulted. Though conditions may be suitable for launching, surf conditions at the landing site must also be considered. Sea state and its impact on flight deck conditions may affect launching helicopters. Wind is also a prime consideration. Winds of 20-25 knots will begin to restrict helicopter operations.

The capability of AAVs, RRCs/CRRCs, and helicopters to reach shore and return without refueling will be affected by distance, sea state, currents, tide, and accurate navigation. If long distances are involved, the use of intermediate platforms such as landing craft, utility (LCUs), and LCACs may be considered. Forward arming and refueling points and small deck amphibious ships may be used to provide longer legs to helicopters. Intermediate platforms to transport RRCs/CRRCs to a launch point can be of great value in assisting in navigational problems, especially in over-the-horizon operations.

Navigational aids (NAVAIDS) such as global positioning system (GPS) and long-range aid to navigation (loran) systems assist in navigational accuracy during transit. Consider inserting reconnaissance elements to provide terminal guidance to both surface craft and helicopters.

Consider the need for LCUs/LCACs/rigid hull inflatible boats (RHIBs) for intermediate transportation or to assist in navigation. Noise signa-tures of craft must be taken into account (compromise/surprise, need for mobility ashore).

Formation for landing must support scheme of maneuver ashore.

Determine requirements for initial terminal guidance.

Determine Go/No-Go Criteria

Consider the weather, the enemy threat, and equipment failure.

Rehearsals

Rehearsals must be conducted to ensure the plan is understood by all. They validate the plan and provide the basis for plan modification. Rehearsals also test and validate communication plans.

Tactical Deception

Consider assets available. Coordinate the deception plan with higher and adjacent forces. Consider the external deception effort. The tactical deception plan must not contradict the overall deception effort.

Reserve Forces

The reserve, when designated, is an integral part of the raid force. It must be designated early, included in all planning, and exercised in each rehearsal. Reserve rehearsals should include tasks required for each potential contingency.

Consider the capabilities required to execute contingencies. Can the raid force accomplish the mission and deal with potential contingencies without assistance?

Consider the reserve's location. Ensure adequate lift is available to rapidly introduce the reserve if embarked. Response requirements and available lift may dictate the location.

Medical Evacuation Plan

This plan should include the killed in action evacuation plan. Include medical support requirements at the objective. Determine medical evacuation points for serious and non-walking wounded. Also determine primary and alternate means of evacuation.

Rules of Engagement

The Law of War applies to raids. Identify other restrictions in effect. Each member of the raid force must understand rules of engagement and how to enforce them. Request relief from those determined to hinder mission accomplishment.

Communications

Identify communication needs (ultrahigh frequency, very high frequency, high frequency, and satellite communications).

Identify critical communications nets and the communications contingency plan.

Consider connectivity between raid force elements and EMCON.

Develop a no-comm plan, brevity codes, and an execution checklist.

Fire Support

Establish fire support coordination center and/or supporting arms coordination center. Coordinate timing of air, naval surface fire support, and other supporting arms throughout the operation.

Determine the need for organic air defense within the raid force.

Determine request and initiation of fire support procedures.

Approach to the Objective

Designate routes so they are easily identified. Properly employ security elements. Maintain communication discipline. Conduct final preparations at the objective rally point. Leaders conduct personal reconnaissance.

Assault on the Objective

Security elements isolate the objective. Support and assault elements move into assault and firing positions. The raid force commander initiates attack of the target. The target is attacked by assault elements.

Withdrawal

Accomplish withdrawal as quickly as possible. Mutual recognition procedures and rehearsals are key to rapid withdrawal.

All should know routes and the sequence of withdrawal. Routes must be easily identified. Incorporate alternate withdrawal routes and methods in the plan.

Account for all raid force personnel.

Excecute fires in support of the withdrawal.

Develop E&E plans.

Appendix D
Considerations for Helicopterborne Raids

A helicopterborne operation is a tactical operation normally in support of the ground tactical plan where troops, supplies, and/or equipment are moved by helicopters. Helicopterborne operations allow the commander to maneuver rapidly to achieve tactical surprise and mass his forces, regardless of terrain obstacles and without depending on ground lines of communication. The unique versatility, strength, speed, agility, and firepower of a helicopter allows the raid force to rapidly maneuver against the raid target and quickly withdraw from the objective area.

CAPABILITIES

A helicopterborne raid force provides commanders with truly unique capabilities. Helicopters help extend the battlefield, move forces, and rapidly concentrate combat power while imposing minimum fatigue on raid force members. Helicopterborne operations are not merely movements of Marines, weapons, and materiel by helicopter units. They are deliberate, precisely planned, and vigorously executed combat operations designed to allow friendly forces to strike over extended distances and terrain barriers to attack the enemy when and where he is most vulnerable. These same attributes allow the formation of tactically tailored helicopterborne task forces that can be employed across the operational continuum. Specifically, helicopterborne task forces--

- Attack enemy positions from any direction.
- Overfly or bypass barriers and obstacles and strike

objectives in otherwise inaccessible areas.

- Conduct raids beyond the forward line of own troops or point of contact using helicopters to insert and extract forces.

- Rapidly concentrate, disperse, or redeploy to extend the area of influence.

- Provide a responsive raid force reserve allowing the commander to commit a larger portion of the force to action.

- Rapidly place forces at tactically decisive points in the objective area.

- Provide surveillance or security over a wide area.

- Rapidly secure and defend key terrain such as crossing sites, road junctions, or bridges.

- Bypass enemy positions.

- Achieve surprise.

LIMITATIONS

A helicopterborne raid force will be relatively light, mobile, and may rely on helicopter support throughout the operation. Initial elements are often separated from weapon systems, equipment, and materiel that provide protection and survivability on the battlefield. They may be limited by--

- Severe weather, extreme heat and cold, and other environmental conditions. Blowing snow and sand limit flight operations or helicopter lifting capability.

- Dependence on air lines of communication.
- Enemy aircraft, air defense, electronic warfare action, and ground attacks by infantry, artillery, and armor. Helicopterborne forces are particularly vulnerable during embarkation and debarkation phases.
- Reduced ground mobility and vehicle-mounted antitank systems once inserted.
- Availability of suitable LZs or helicopter rope suspension sites.
- Battlefield obscuration that limits helicopter flight.
- High fuel and ammunition consumption rates.
- Nuclear, biological, and chemical protection and decontamination assets.

SEQUENCE

Successful use of helicopters requires a careful analysis of METT-T and detailed reverse planning. Five basic plans that comprise the reverse planning sequence are developed for each helicopterborne operation. They are the--

- Ground tactical plan.
- Landing plan.
- Air movement plan.

- Loading plan.

- Staging plan.

These plans should not be developed independently. They are coordinated and developed concurrently by the staff of the designated helicopterborne raid force. Sample formats for each plan are provided in FMFM 6-21, ***Tactical Fundamentals of Helicopterborne Operations, appendix J.***

THE GROUND TACTICAL PLAN

All planning evolves around the ground tactical plan. It specifies actions in the objective area which ultimately accomplish the mission. It may also include subsequent operations. The ground tactical plan for the helicopterborne raid contains essentially the same elements as any other raid. Infantry elements are placed on or near the objective and task-organized for immediate attack of the target. In some situations, the helicopterborne raid force cannot be introduced quickly into the objective area and must land away from the target and conduct a foot and/or vehicle movement to the objective. The scheme of maneuver may assume a variety of possibilities depending on the commander's evaluation of METT-T. The availability of LZs in the area is very important.

Mission

The most obvious portion of the helicopterborne mission is movement. The following concerns must be addressed:

- Missions of all task force elements and methods for employment.

- Purpose of the helicopter movement.

- Number of Marines to be lifted.
- Total weight to be lifted.
- Internal and external lift requirements.
- Distance of the air movement.

Concept of Operations

The ground concept is formulated in five parts:

- Ground movement to the pick-up zone.
- Securing and organizing the pick-up zone.
- Actions at the insert LZ.

- Ground movement from the LZ to the objective area or actions on the objective if the force is landed directly on the target.

- Operations subsequent to securing the objective.

THE LANDING PLAN

The landing plan must support the ground tactical plan. The landing plan sequences introduction of elements into the objective area so that units arrive at locations and times prepared to execute the ground tactical plan. The landing plan should include--

- The availability, location, size, and enemy proximity to potential LZs. The helicopterborne force is most vulnerable during landing. Elements must land with tactical integrity. Troops are disoriented if the briefed landing direction changes and they are not kept informed. The helicopterborne raid force must land prepared to fight in any direction.

- Supporting fires (artillery, naval surface fire support, and CAS) to support insert and extract.

- Provisions for emergency extract, resupply, and medical evacuation.

Selecting LZs

The raid force commander selects the LZ based on recommendations by the air mission commander. LZs are selected based on--

- The raid force commander's concept of operations.

- If LZs can be located on, near, or away from the objective, depending on METT-T.

- LZ size. Size will determine landing sequence. It is desirable to land the entire force simultaneously to minimize exposure to the threat. The size of available zones may also require using more than one zone, more than one wave, or increased separation between waves. Weather, reduced visibility, or strong winds may preclude or limit the use of marginal LZs. LZs should be readily identifiable from the air.

- Alternate LZs. Alternates should be planned for each primary LZ to ensure flexibility.

- Enemy troop concentrations, air defenses, and the enemy's capability to react to a helicopterborne landing.

- Sites that deny enemy observation and acquisition of friendly ground and aviation elements while they are en route to, from, and in the LZ.

- Requirements for logistic support and fire support.

- Routes to and from LZs.

Single Versus Multiple LZs

In addition to deciding where to land in relation to the objective, a decision must be made on whether to use single or multiple LZs.

Single LZs allow concentration of combat power in one location and facilitate control, orientation, and supporting arms coordination. They enhance security for subsequent lifts, reduce route planning require-ments, centralize required resupply operations, and require less planning and rehearsal time.

In multiple LZs, the force is dispersed, thus avoiding creation of a lucrative target for enemy mortars, artillery, and CAS. Multiple LZs allow rapid employment of elements to accomplish tasks in geographically separated areas. They reduce the enemy's ability to determine where to employ his main effort and react to the initial lift if force is detected. Multiple LZs eliminate aircraft congestion and make it difficult for the enemy to determine the helicopterborne raid force's size and the exact location of supporting weapons.

Fire Support

Helicopters should land without preparatory fires to achieve tactical surprise. However, fires are planned to support a landing in each LZ if they are needed. When developing fire support plans, consider--

- Deception. False preparations are fired into areas other than the objective or LZ.

- Loss of surprise.

- Availability and type of fire support.

- Significant targets. A known or suspected enemy force regardless of size warrants targeting.

- Obstacles to landing and maneuver. Artillery, bombs, or napalm can cause craters, tree blowdown, fires, and LZ obscuration.

- Scheduled fires.

- Positive control measures. Control measures must be established for lifting or shifting fires. Airspace coordination areas may be necessary to protect approach and retirement lanes.

THE AIR MOVEMENT PLAN

The air movement plan is based on the ground tactical and landing plans. The plan provides the air movement schedule for troops, equipment, and supplies. It also provides coordinating instructions on approach and retirement routes, air control points, aircraft speeds, altitudes and formations. The use of attack helicopters should be included in this plan. The air movement plan is approved by the MAGTF commander based on recommendations from the air mission and raid force commanders.

Evaluation of METT-T (with strong emphasis on threat analysis) determines approach and retirement route selection. Route and altitude are interdependent and should be considered concurrently to determine optimum movement. Avoidance of enemy detection and fires is the primary consideration. Consider the following in route selection:

- Use terrain to the best tactical advantage.

- Identify primary and alternate approach and retirement routes.

- Select routes that are easy to identify and navigate.

- Communications must not be easily impaired.

- Routes can be flown under adverse weather conditions.

- Unique or complicated support requirements are minimized.

- Supporting arms capabilities are maximized and limitations are minimized.

Ground commanders must know the dimensions of approach and retirement routes for fire support planning. Leaders must maintain situational awareness including navigation during flight as helicopters may be forced to land unexpectedly.

Planned fires in support of the raid may be executed during flight. These fires are planned as would normally support any helicopterborne operation. Fires along the flight route are planned to suppress known or suspected enemy positions. These fires should be of short duration. Multiple target engagement techniques should be used.

Fire plans must be complete and flexible. Fire support plans include suppression of enemy air defense systems and smoke to protect formations from enemy detection from pick-up zone, along approach and retirement lanes, and LZs.

On-call fires are planned along the flight route for rapid adjustment of targets of opportunity. Using illumination requires detailed planning as it effects security and use of night vision goggles.

THE LOADING PLAN

The loading plan is based on the air movement plan. Correct helicopter loading is essential in maintaining proper employment and orientation upon debarkation. In turn, proper debarkation enhances mobility and reduces the time to organize for movement from the LZ to the target. The key is to load and land helicopters in the manner and sequence which allows immediate assumption of the mission upon landing.

The loading plan identifies the Marines, equipment, and supplies to be loaded on each aircraft. Each helicopter load is prioritized to establish the raid force bump plan. A bump sequence is designated so essential Marines and equipment are loaded ahead of less critical loads in case of aircraft breakdown or other problems. Likewise, each individual heliteam establishes an internal prioritization should conditions preclude the loading of the entire team. Planning must cover the organization and operation of the pick-up zone including load positions, day and night markings, and communications. The loading plan becomes more complicated when mixing internal and external loads and/or when mixing helicopter types.

Detailed, written plans can be reduced by having adequate unit SOPs covering pick-up zone tasks and loading plans. During preparation of loading tables, unit leaders at all levels attempt to maintain tactical integrity of units. Fire teams and squads load intact on the same aircraft and platoons in the same wave. The goal is to ensure unit integrity is maintained at every level.

Every towed item is accompanied by its prime mover. Crews are loaded with their vehicle or weapon and the appropriate ammunition. Sufficient personnel are on board to unload cargo. Communication between flights is maintained. Loads should be planned so that leaders, special equipment, crew-served weapons, or a capability are not lost with the loss or destruction of one aircraft.

THE STAGING PLAN

The staging plan is based on the loading plan and prescribes the arrival time of ground units (troops, equipment, and supplies) at the pick-up zone in the proper order for movement. Loads must be ready before aircraft arrive. The staging plan also provides for zone organization, defines flight routes to the zone, and provides instructions for linkup of all aviation elements. Air-to-air linkup of aviation units should be avoided, especially at night when night vision goggles are being used.

THE MISSION BRIEF

The mission brief is the final phase of planning and should include all key personnel. Responsibility for operational briefings is a function of command and rests with the commander tasked with executing the helicopterborne operation. This brief will set forth the concept of operations, scheme of maneuver, fire support plan, and specific details on mission coordination and execution.

Mission Briefing Guide

Information shall be provided which enables each participant to understand the overall operation and his specific role and responsibilities regarding mission execution. Joint briefings with representatives from each participating unit should be conducted. It must be emphasized that the mission briefing guide is *only* a guide. Include only those items that apply directly to a specific mission. Since all members of the mission will not be involved in planning, the flight brief must be well-delivered, organized, and easily understood. Tactical SOPs will significantly reduce the time required. The most common problems associated with mission briefings are--

- Redundancy.

- Air mission commander (AMC) is overtasked.
- Communication plan has different frequencies than those in use.
- AMC position in-flight not briefed.
- Inadequate or improper use of flight coordinator.
- LZs are not thoroughly briefed.
- Inadequate scatter plan.
- Attack helicopter positions during flight sequence not briefed.
- Failure to synchronize time with all personnel. At a minimum, attendees should include the--
- Air mission commander.
- Raid force commander.
- Fire support coordinator.
- Tactical air controller (airborne) or forward air controller (airborne).
- FAC and appropriate level air officers.
- Fixed-wing attack aircraft flight leader.
- Attack helicopter flight leader/flight coordinator.
- Helicopter transport commander.

- Helicopter aircraft commanders.

Mission Brief Guide Format

This format will help the raid force commander ensure essential information is included in helicopterborne mission briefings.

Situation

Enemy forces (especially troop concentrations and locations and types of enemy antiaircraft weapons).

Weather (ceiling, visibility, wind, temperature, pressure and density altitude, sunrise and sunset, moonrise and moonset, percent of moon illumination, end of evening nautical twilight, beginning morning nautical twilight, pick-up zone and LZ altitudes, and weather outlook).

Friendly forces.

Mission

Clear, concise statement of the task that is to be accomplished (who, what, when, where, and why).

Execution

Commander's intent.

Ground tactical plan.

Fire support plan to include suppression of enemy air defenses.

Air defense plans.

Engineer support plan.

Fixed-wing air support.

Aviation unit tasks.

Staging plan (both primary and alternate pick-up zones).

- Pick-up zone location.
- Pick-up zone time.
- Pick-up zone security.
- Flight route to pick-up zone.
- Pick-up zone marking and control.
- Landing formation and direction.
- Attack and air reconnaissance helicopter linkup with lift elements.
- Troop and equipment load.

Air movement plan:

- Primary and alternate flight routes.
- Initial points.
- Flight formation(s) and airspeeds.
- Deception measures.
- Air reconnaissance and attack helicopter missions.

- Abort criteria.
- Helicopter wave and serial assignment table (if used).

Landing plan (both primary and alternate LZs):

- Landing zone location(s).
- L-hour.
- Landing formation and direction.
- Landing zone marking and control.
- Air reconnaissance and attack helicopter missions.

Extraction plan (both primary and alternate pick-up zones):

- Pick-up location.
- Pick-up time.
- Air reconnaissance and attack helicopter missions.
- Supporting plans.

Return air movement plan:

- Primary and alternate flight routes.
- Initial points.
- Flight formations and airspeed.
- Air reconnaissance and attack helicopter missions.

- Landing zone locations.
- Landing zone landing formation and direction.
- Landing zone marking and control.

Coordinating instructions:

- Go/no-go criteria.
- Mission abort.
- Downed aircraft procedures.
- Weather decision by on-hour increments and weather abort time.
- Passenger briefing.

Service Support

Forward arming and refueling point(s) locations (primary and alternate) and procedures.

Ammunition and fuel requirements.

Backup aircraft.

Aircraft special equipment requirements.

Medical service support.

Command and Signal

Command:

- Location of raid force commander.

- Point where air reconnaissance and attack helicopters come under operational control of ground maneuver elements.

Signal:

- Radio nets, frequencies, and call signs.

- Communications-electronics operating instructions in effect and time of change.

- Challenge and password.

- Authentication table in effect.

- Visual signals.

- NAVAIDS (frequencies, locations, and operational times).

- Identification friend or foe.

- Codewords.

Time Hack

All watches are synchronized.

Briefing the Flight Crew

Proper briefing of flight crews is essential to mission success. The mission briefing should be conducted in the most logical, concise, and organized manner possible. Seat flight crews together by section, division, or flight. Provide preprinted sheets/cards with nonperishable administrative information.

Essential subjects to brief on are--

- Helicopter assignment.
- Call signs.
- Flight leader/alternate flight leader.
- Times (manning, engine start, rotor engagement, taxi, or takeoff).
- Controlling agencies.
- Frequencies.
- Radio procedures.
- Identification friend or foe procedures/codes.
- NAVAIDS.
- Weather forecast.
- Minimum operational weather.

Appendix E
Considerations for Mechanized Raids

A mechanized operation is a tactical operation designed to maximize ground mobility, protection, shock action, and firepower of the force to concentrate combat power rapidly against the enemy. Planning mechanized raids covers those concerns found in planning any raid operation and those inherent in a conventional mechanized operation. Mechanized assets available to the raid force may merely provide transportation to and from the objective or be an integral part of objective area actions.

CAPABILITIES

A mechanized force is a task-organized ground combat force of combined arms built around an infantry or tank unit and reinforced with substantial assault amphibian or light armor assets. Normally supported by air, artillery, naval surface fire support, antitank, and engineer units, mechanized raid forces employed with assault amphibians also conduct amphibious raids.

Capabilities of mechanized and armored vehicles provide the force with a number of employment options. Protection against fragmentation and small arms lends itself to evacuation operations. Speed and firepower lend themselves to employment in destruction raids and diversions. Even small but violent incursions with mechanized raid forces can result in enemy overreaction, create doubt, and cause him to change his plans. During peacetime operations, the presence of mechanized forces can subdue or thwart potential actions by insurgent factions or other extremist groups.

Firepower

The M-1A1 tank is equipped with the 120mm high velocity main gun, a turret-mounted .50 caliber machine gun, the loader's 7.62mm machine gun, and a coaxially-mounted 7.62mm machine gun. Assault amphibians are equipped with cupola-mounted .50 caliber and 40mm machine guns. The light armored vehicle family of vehicles is equipped with the 25mm chain gun; 7.62mm machine guns; the tube launched, optically tracked, wire command link, guided missile system (TOW); and the 81mm mortar, depending on the variants.

Armor Protection

Tanks, assault amphibious vehicles, and light-armored vehicles have sufficient armor to withstand shell fragmentation and, depending on range, small arms fire.

Mobility

Tanks, assault amphibious vehicles, and light-armored vehicles have a cross-country capability which permits rapid concentration and mobility. The amphibious capability of the assault amphibious vehicle also allows the raid force to enter and exit the objective area from the sea when landing sites permit.

Shock Action

Firepower, armor protection, and mobility of the mechanized force, when properly employed, provide shock action which can paralyze the enemy and hinder his ability to react to the attack.

LIMITATIONS

Size, weight, noise, and limited observation are inherent limitations. Deep mud, rocky or stumpy ground, dense woods, snow, and swamps limit their use. Rolling terrain (where cross-country mobility can be exploited) is best suited for mechanized operations. Manmade obstacles canalize movement and may restrict employment. Obstacles frequently encountered are tank ditches, tank traps, and antitank minefields. Urban areas are also considered obstacles.

MOVEMENT

Movement techniques are those normally employed in most mechanized operations. The relatively large signature projected by a mechanized force may require extensive deception and well-planned routes to take advantage of terrain masking and ambush avoidance. Infantry or reconnaissance units and aviation assets may provide ground surveillance and security along designated routes in restrictive terrain. Particular attention must be given to passage and penetration of friendly and enemy obstacle belts.

ACTIONS AT THE OBJECTIVE

Plans may require direct employment of mechanized assets and their associated weapons at the target. In such instances, the force is task-organized and controlled in the same manner as for any other raid. Factors determining mission assignment are vehicle type, its capabilities, and its limitations.

A battalion-sized raid force (based on a notional battalion landing team comprised of tanks, light-armored vehicles, and assault amphibians) assigned a destruction raid may be task-organized as follows:

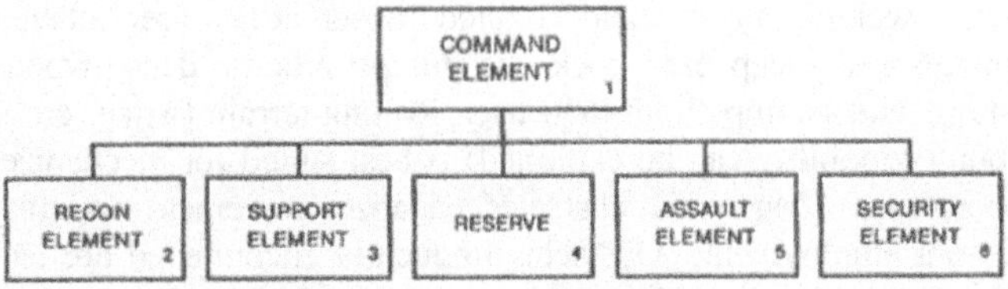

1 - 2 AAVC-7s/LAV C^2

2 - 1 LAR company

3 - 1 heavy machine gun platoon (HMMWV)
1 antiarmor (TOW) platoon/section (HMMWV)

4 - 1 helicopterborne rifle company
2 AH-1Ws
1 section F/A-18/AV-8B

5 - 1 tank platoon
1 AAVP-7 mounted rifle company

6 - 1 LAI company

(reverse blank)

Appendix F
Considerations for Small Boat Raids

Military and paramilitary units have successfully conducted small boat insertions and extractions for a variety of raiding purposes. MEUs have employed small boats routinely since the introduction of the CRRC and RRC within the MEU table of equipment and the internal designation of a small boat company. Small surface craft maximize the advantage of surprise through stealth as the relative signature of small raiding parties embarked aboard small boats is minimal.

CRAFT DESCRIPTIONS

The Zodiac Marine Commando F470 (or CRRC) is a 15-foot, 5-inch inflatable rubber boat with a rigid aluminum deck. The aluminum deck is composed of four lightweight, self-locking aluminum sections and two aluminum stringers. *Locally-machined solid aluminum stringers are required for moderate to heavy surf conditions.* The CRRC can mount a single 35 to 55 horsepower (hp) engine or twin 35hp engines on the reinforced transom. It can be launched from various platforms at sea including LCU. It has low visual and electronic signatures.

The Boston Whaler Rigid Raider (or RRC), is 18 feet 6 inches long with a fiberglass hull and two 70hp outboard engines. Its large size and weight make it extremely dangerous to handle in surf conditions, but performs extremely well in protected waterways. It has high speed, good maneuverability, and long range. In the RRC, a MAGTF has a high speed, surface, low signature, amphibious method for insertion and extraction. One limitation of the RRC is that it requires a launch and recovery device (LRD) mounted on the stern gate of an LPD amphibious assault ship.

NOTIONAL COMBAT RUBBER RAIDING CRAFT/RIGID RAIDING CRAFT RAID FORCE

Each mission assigned to the CRRC/RRC company will require a specific task organization based on METT-T analysis and logistics (such as fuel required). The notional raid force shown is provided for illus-tration only and based on the availability of 10 CRRCs/RRCs.

Though the CRRCs/RRCs manufacturer advertises a 10-man load capability, optimum weight limit is 2,000 pounds. Smaller boat teams still capable of manipulating the craft without mechanical power (normally six to eight), allow for wounded and enemy prisioner of war transportation, and increased flexibility for dealing with bump plans, sea states, and equipment requirements. Commanders may desire to increase or decrease the size of boat teams based on the tactical situation.

The notional organization is intended to allow as much unit integrity as possible. While each element might include Marines from several different units, it is essentially structured around the company's normal team, squad, and platoon assignments. This allows the teamwork established during the company's individual and team training to carry over into the raid force organization. Substitutions, deletions, or additions can be made. Flexibility is the key.

Command Element	Officer	Enlisted	Navy
CO	1		
CO GySgt		1	
RTO (CO TAC)		1	
FAC	1		
RTO (Bn TAC)		1	
Corpsman			1
Coxswain		1	-
	2	4	1
Assault Element			
Team Leaders	2		
Assistant Team Leaders		2	
RTOs		2	
2 SMAW Teams		2	
4 Fire Teams (2 each Aslt Team)		16	
2 Demo Teams		4	
	2	26	0
Support Element			
Element Leader	1		
3 MG Teams		9	
Fire Team		4	
SMAW Team		1	
Corpsman			1
	1	14	1
Recon/Security Element			
Element Leader	1		
3 Security Teams		17	
	1	17	0
TOTAL RAID FORCE	6	61	2

PERSONNEL ASSIGNMENTS

A number of Marines serving in rifle companies employing the CRRC/RRC assume responsibilities unique to that company.

Coxswains

Each boat is assigned a primary and alternate coxswain. Certain situations will require coxswains to remain with boats at the landing site or to move the boats to a different location for extraction of the force. Primary coxswains should not be considered available for actions at the objective. Principal duties of the coxswain include responsibility for--

- Their assigned CRRC/RRC, engine and equipment maintenance, and operational checks before and after each operation. They assist the boat team leader with all CRRC/RRC related matters.

- Tactical operation of their assigned CRRC/RRC and knowledge of all signals, formations, and procedures for CRRC/RRC operations.

- Control of personnel assigned to their raiding craft.

Scout Swimmers

CRRC/RRC company landings are normally preceded by operations involving reconnaissance units/sea-air-land teams.

Both forces are normally assigned the task of securing the landing site and providing initial terminal guidance. However, operational requirements may preclude availability of these units. On these occasions, the CRRC/RRC company must make its own beach reconnaissance before landing the entire raid force. One method by which the company may do this is by scout swimmers.

Scout swimmers are responsible for beach reconnaissance before the CRRC/RRC enters the surf zone. They are first class swimmers or better, are proficient in the conduct of beach reconnaissance, and can maintain a course heading while swimming at night using a compass.

Boat Teams

Each CRRC/RRC contains a 7- to 8-man boat team. The team consists of six Marines, a coxswain, and when required, an assistant coxswain/navigator. If an assistant coxswain/navigator is designated, he may be colocated with the coxswain at the stern. The team is numbered consecutively from bow to stern, odd numbers to starboard, even numbers to port.

The number one man is the strokeman. He sets the rate of paddling and maintains paddling rhythm as directed by the coxswain. He assists the coxswain in keeping the boat perpendicular to the breaker line when beaching or launching the boat. He assists the coxswain in avoiding obstacles in the water.

The number two man also assists the coxswain in keeping the boat perpendicular to the breaker line and in avoiding obstacles. He handles the tow line during towing operations.

Numbers three and four are responsible for lashing and unloading equipment in the boat. They also serve as scout swimmers if scout swimmers are not carried as passengers.

Numbers five and six assist in lashing and unloading equipment in the aft section of the boat. They assist the coxswain in maneuvering in swift currents and rig and handle the sea anchor, when used.

Number seven (assistant coxswain) and number eight are responsible for the performance of the crew, the handling of the boat, and for the internal distribution of men and equipment. The coxswain issues all commands to the boat team, maintains course and speed, and operates the engine when used.

LAUNCHING

CRRC Launch

There are three recommended methods for launching the CRRC from the LPD, LSD, or LST stern gates. Ships should be underway and heading into the seas, and be ballasted so that the bottom of the stern gate is in contact with the water's surface. Underway speeds may vary from steerage to greater speeds although sea conditions at the stern gate will dictate the speed for launch. Generally, the speed may be between 3-5 knots.

Bow First. Up to three craft can be positioned on the stern gate of an LPD or LSD or up to two on an LST. Craft are positioned with their bows aft and paddles are placed against the traction bars, ready to push.

Coxswains signal the stern gate noncommissioned officer (NCO) who will visually verify readiness. He notifies the well deck control officer that the gate is ready to be lowered. As the gate is lowered, the craft begin to float and the boat teams push off.

As each CRRC clears the ramp area, the coxswain directs the lowering and starting of the engine or does so himself. Boats follow 50 to 100 meters behind the ship in staggered column or line astern formation until all craft are launched.

Stern First. CRRCs are placed on the stern gate so that the transoms are even with the outside edge of the stern gate and the engine is lowered. As the stern gate is lowered and the CRRCs begin to float, boat teams back paddle off the gate. With the engine in the water, it can be quickly started. Stern first launch allows the coxswain to--

- Maintain positive control of the CRRC.
- Keep eye contact with safety personnel.
- Operate engines before launch.
- Control launch speed.

Bow First with Sea Anchor. This method is executed in the same manner as bow first, except that each CRRC will have a sea anchor on approximately 30 feet of line. The anchor is thrown into the sea and must be fully deployed before the stern gate is lowered. The stern gate is lowered and the CRRC is pulled by the anchor from the ship. This evolution can be very quick and rough for the CRRC. This option has serious safety considerations and should be a last resort for launching the raid force.

LCU Launch

CRRCs can be launched from an LCU acting as an intermediate transport vessel using the bow ramp. The LCU craftmaster will conduct a brief to discuss correct procedures.

The LCU is stopped with its stern to the sea. Sea state is not greater than 1 or as determined by the LCU craftmaster. Only one CRRC is launched at a time.

The bow ramp is lowered to the horizontal position. One craft is positioned with its stern toward the end of the ramp keeping the CRRC engine clear of the bow ramp and over the water. The boat team boards the craft.

The coxswain notifies the ramp NCO who will visually verify

readiness of the craft. The ramp NCO notifies the bow ramp operator to lower the bow ramp. The ramp is lowered into the water with the block resting on the stops to keep the stress off the bow ramp wires.

With the ramp in the water, the boat team pushes off the ramp with their paddles and then backpaddle. The engine is lowered and started when the craft has fully cleared the bow ramp. When clear, the bow ramp operator raises the ramp back to the horizontal position.

Hoisting

In the absence of a well deck the CRRC can easily be hoisted over the ship's side into the water. The easiest way is to use the hoisting sling provided by the boat manufacturer. Or, a basket can be made from peck and hale nets with plastic pipes to give the basket shape. The basket serves as a sling in which the CRRC rests.

RRC Launch

The RRC's LRD provides a platform for launching and recovering RRCs aboard LPD amphibious ships. It allows for dry well operations without having to ballast down and empty the well deck. The LRD is designed to achieve rapid deployment and recovery of the RRC.

LRDs are assembled on the ship's stern gate while in the horizontal position. The dolly with the RRC is pushed into the LRD stern first. Chocks are installed to hold the dolly in place while personnel enter the boat with their gear.

The stern gate is lowered until the boat propellers are underwater. As the water intake of each engine is submerged, they are started. The stern gate is lowered until the boat floats free of the dolly pads and is backed out of the LRD. The coxswain must keep the RRC aligned while backing out of the trap.

The empty dolly is removed from the LRD after the stern gate is raised to the horizontal position and the wheel chocks removed. This procedure is repeated until all RRC are launched.

FORMATIONS

Formations are critical to the company commander's ability to control his raid force while in open and restricted waters. Each tactical forma-tion has an order of travel. The lead or command boat (number 1) contains the primary or key navigator whose responsibility is to ensure the section or company gets to its correct destination. The second boat contains the alternate or secondary navigator. The four basic movement formations are column, staggered column, wedge, and line.

Column

This is an administrative formation for a small number of craft, usually five or less. A company split into sections can use the column formation.

Staggered Column

This is the easiest formation for the coxswains to maintain visual contact. It also keeps boats from ramming each other if one in the front has engine failure. This formation is used when there is calm water and/or good visibility, and traversing areas with confined maneuvering, such as harbors.

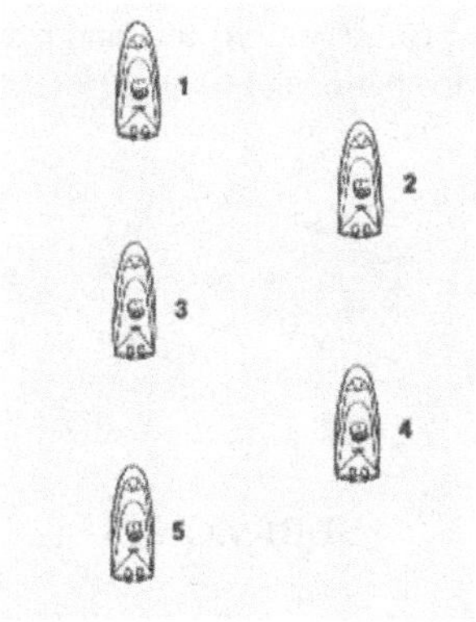

Wedge

The wedge is not as easy to maintain as the column formation, but it allows easier coxswain control because other boats are easier to view. This formation is more difficult to pick up from the air and can provide better security ashore. The wedge is for open water and/or reduced visibility.

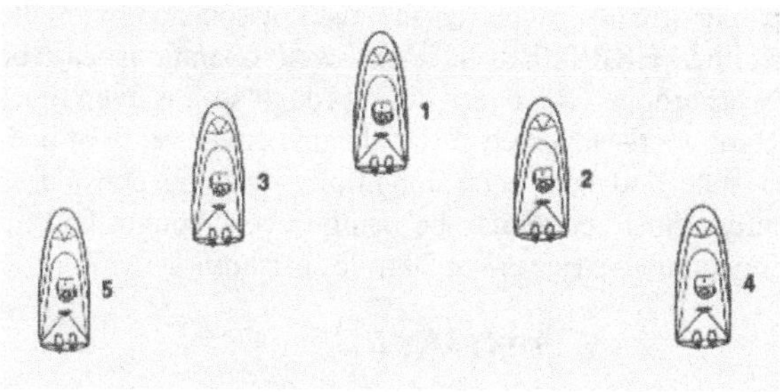

Line. The line is primarily used for landings. It is difficult to maintain for extended periods of time.

SURF ZONES

A key planning consideration for a small boat insert will be the characteristics of the surf zone at the landing site. Although scout swimmers conduct a reconnaissance of the surf zone and the beach, advance information on these conditions will greatly enhance the commander's ability to plan and determine go/no-go criteria.

SEA STATE

Rough weather greatly increases the risks associated with the launch of either CRRCs/RRCs. Wet well operations can be particularly hazardous. Recovery during rough seas is even more difficult. High winds adversely affect the maneuverability of both craft. Sea state and surf zone conditions must be considered independently. Surf zones may be benign even though the raid force may encounter sea state 3 at 20 miles offshore.

DISTANCE

Generally, the initial point in over-the-horizon operations is considered anything out of visual range or coastal radar range of the shore. This distance may be substantially greater against a sophisticated enemy with extended detection capability or offshore facilities such as gas/oil platforms in the target area. The commander must also consider sea states, time required to launch and transit to the objective on time, and weather.

NAVIGATION

Navigation involves the movement of a vessel expeditiously and safely from one point to another. Both art and science are involved in navigating a craft accurately to a beach or rendezvous point at sea. The science of navigation includes the mathematical computation of navigational problems and the use of tables, almanacs, and publications to increase accuracy. The art of navigation involves the proficient use of these NAVAIDS and the different navigation methods with good judgement, prudent application, and common sense. There are four types of over-the-horizon navigation:

- Radio/electronic.
- Dead reckoning.
- Piloting.
- Celestial.

In keeping with EMCON conditions, the preferred navigational method for a raid force is nonelectronic. Thus, dead reckoning and coastal piloting are the primary methods of navigation. Passive electronic methods (GPS, loran, or satellite navigation) are useful complements, but they should not be totally relied on due to problems in propaga-tion, signals interpretation in adverse weather conditions, and the delicacy of electronic instruments.

Radio/Electronic

GPS. An all-weather, worldwide satellite system that gives a three-dimensional fix within 16 meter accuracy.

Loran. Loran is an all-weather coastal navigation system covering two-thirds of the U.S. and limited areas of the world. It consists of 44 master and secondary (slave) stations that transmit on a continuous basis. Loran is designed for military tactical use using transportable transmitting stations.

Dead Reckoning

Dead reckoning is the basis of all marine navigation. It is the only method by which a craft's estimated position can be determined by applying the course traveled, speed of the craft, and time run from the last known position. Dead reckoning is used to--

- Construct the navigation plan.
- Estimate position at sea using time, distance, and speed, while enroute.
- Establish a new route, after course and speed changes or a position update.

Piloting

Piloting involves frequent and continuous determination of position by reference to landmarks and NAVAIDs, then adjusting the craft's heading to correct errors. The most important aspect of piloting is *preparation.* Study charts and publications to become familiar with shoals, tides, currents, and NAVAIDs in the raid area.

Minimum equipment required for accurate boat navigation is a boat compass (nautical compass), a laminated chart with plotting board, a speed determining device, a time piece, a nautical slide rule, and a hand bearing compass.

Celestial

This is the oldest form of navigation. It uses observations of the Sun, Moon, stars, and planets in relation to the Earth to determine position. Extremely weather-dependent, it must have good conditions and takes a great deal of training.

Navigation Duties

Navigator. The navigator constructs the navigation plan, oversees its execution, and is charged with the safe piloting of the craft. He orders course and speed changes to the coxswain and keeps the total time of passage. He also operates the plotting board and notifies the assistant navigator when to take bearings and speed checks.

Assistant Navigator. The assistant navigator assists the navigator as directed. He takes bearings, measures speed, records time between legs, and stays alert to observe sea conditions, aids to navigation, and land features.

Nontactical Navigation Considerations

After collecting all available information, the navigator will ascertain the type and location of available NAVAIDs. The navigator should know all he can about the following:

Weather. The forecasted weather (available from the ship's navigator or meteorological detachment) should be obtained in as much detail as possible. Sea conditions can change rapidly. Weather should be considered not only at the time of launch, but throughout the raid to ensure the safe return of the raid force. Cloud cover and astronomical data and their effects on the raid force should also be considered.

Ocean Current. The estimated and actual ocean current is critical to the conduct of current sailing. Allowances must be made for the effects of an estimated or actual current to determine what course and speed the craft must make good through the water so that in the end, the intended track and then actual track will coincide. The estimated current is used in planning until the actual current is determined underway.

Tidal Current. If the raid is taking place inside a large bay or up a navigable river, tidal currents must be studied since unexpected currents may delay the raid force.

Tidal Range. If the raid is taking place in an area affected by tides, failure to consider tidal range may result in the raid craft coming ashore in an area where there is not enough water.

Beach Type. The type of beach at the beach landing site is important to the success of the raid. Landing in rocky areas or along a cliff requires special training and preparation. Information on beach types can be gained from the G-2/S-2 or from real-time reports from reconnaissance elements.

NAVAIDs. The navigator must be familiar with all the available NAVAIDs in the vicinity of the raid to assist in positioning once in navigable waters. He should especially look for land forms, lighthouses, towers, lights, buoys, horns or other sound devices.

Navigation Hazards. Any sunken vessels, shoals, sandbars, or rocks that may impede the movement of the raid force to the beach landing site.

Local Surf Conditions. These include wave heights, period and type, and any known unusual conditions.

Tactical Considerations

Certain tactical considerations are important to the navigator as he develops the navigation plan. When selecting the insert point, rendezvous point, and beach landing site, the navigator not only considers the NAVAIDs, but the enemy situation as well. Key considerations are--

- Location, type, and range of enemy sensors, alarms, or radar.
- Patrols near the beach landing site.
- Schedules and routes of patrol craft.
- Schedules, routes, and capabilities of air patrols.
- Wire, mines, or lights near the beach landing site.
- Alert status of units near the beach landing site.
- Offshore platforms and anchorages that may interfere with the route.
- Capability to reinforce the area of the beach landing site.
- Location, type, and range of antiship missiles that may interfere with the ship, an insert point, or a rendzvous point.
- Location of possible cache points and emergency helicopter LZs.
- Beach exits for rapid departure once ashore.

SAFETY

Safety is critical during CRRC/RRC operations as small boat operations are inherently dangerous. Corpsman are readily available and thoughtfully distributed throughout the force.

Craft are organized into two-craft sections. These craft operate under the buddy system and will always maintain visual contact with each other to provide immediate assistance in an emergency. Boat team members will use the buddy system.

Life vests are worn by all hands. Vests are checked before each launch. Unqualified swimmers will not participate in CRRC/RRC training or operations. Each Marine is equipped with a day/night flare for emergency use (chemlights may be substituted). Only lightweight, easily jettisoned gear is worn. Radios, packs, and other equipment are lashed in the craft. Exposure suits will be worn when conditions dictate.

If a man falls overboard, the first man observing this calls out, "man overboard port/starboard side." The coxswain immediately attempts to maneuver his craft to pick the man up. Only as a last resort will additional men go in the water to provide assistance.

As a last resort, an overboard boat team member discards any equipment necessary to stay afloat. He attempts to maintain visual and sound contact with his craft.

If a craft is broached, the coxswain orders a long count, and buddies pair off. Every effort is made to right the craft or stay with the capsized craft.

MAINTENANCE AND LOGISTICS

Fuel, oil, spare parts and repair parts are required to keep CRRCs/RRCs operating during training as well when employed. Many items must be maintained at the company level to ensure a high state of readiness. Items necessary for any operation are--

- Fuel.
- 2-cycle oil.
- Chemlights for night operations (infared or colored).
- Waterproof bags.
- Nylon rope to run through the CRRC/RRC D-rings across the boat to secure all items to the boat and keep them easily accessible.
- Day/night flares for emergency use only (one per man).
- Snaplinks (one per man, four extras per boat at a minimum).
- Life vests (one per man).

Boat and engine maintenance beyond 1st and 2nd echelon is done by the force service support group. The maintenance section must be familiar with all types of repairs. Other items must be procured for general maintenance and storage for parts and equipment. Mainte-nance personnel should--

- Check all valves; ensure they are tight.
- Check for leaks; patch as needed.
- Check boat handles and seams.
- Secure motor and fuel bladder to boat.
- Open all fuel valves.

- Check paddles; six per craft is the minimum.
- Check tow rope.
- Check floor boards.
- Secure chemlights to rear of boat.
- Check Marines for life vest and safety light.
- Check and secure all gear.

Appendix G
Considerations for Artillery Raids

The artillery raid is another means for the commander to apply high intensity, precise offensive power at the most advantageous time and location. The artillery raid is a unique option that enables the commander to achieve effects on high value targets outside the range of his surface weapons that cannot be attained by other means, or can be attained only by exposing other weapons systems or personnel to unacceptable risk. This option requires the commander to position critical artillery assets ashore or far forward exposing them to significantly increased threat from enemy action.

Artillery raid operations can be effectively integrated with offensive air support, electronic warfare systems, unmanned aerial vehicles, and/or maneuver to produce a combined arms raid. As part of combined arms raids, artillery can serve as the main effort or perform a supporting role. Integration of combined arms can magnify the advantages and capabilities of each arm while mitigating individual disadvantages and limitations.

WEAPONS

The Marine Corps' standard artillery weapon, the M198 155mm howitzer, is characterized by its extended range, lethality and variety of munitions, and significant mobility/lift requirements. The CH-53E and the Army's CH-47 can safely lift the M198. Notwithstanding trafficability and mobility concerns, the M198's destructive power and range make it the optimum weapon system for attacking raid targets. The M101A1 105mm howitzer is distinguished from the M198 as a much lighter weapon with a higher rate of fire and greater mobility that can be employed quickly by fewer personnel and requires little time to emplace. These distinctions may make the M101A1 more suitable for many raid scenarios, particularly those requiring helicopter insertion. Although 105mm ammunition weighs less per round than 155mm ammunition, a 155mm ammunition load will achieve greater effects than a similar size 105mm ammunition load.

ADVANTAGES

The variety of artillery munitions available, particularly 155mm, offers numerous attack options to achieve desired effects out to ranges of 18,100 meters. The availability of conventional high explosives, improved conventional munitions/dual-purpose improved conventional munitions, scatterable mines, incendiaries (white phospherous), and obscuration smoke enables commanders to effectively engage a variety of target types and sizes. Targets can be engaged out to ranges of 30,000 meters with high explosive-rocket assisted projectiles. By attacking from these extended ranges, the possibility of inadvertently alerting the target itself is diminished. If the target is accurately located, artillery raids may not require "eyes-on-target" except to determine target assessment.

VULNERABILITIES AND LIMITATIONS

Considering its size, means of movement, and firing profile, an artillery raid force generates significant radar, noise, and visual signature. Although not normally detectable by the target itself, an artillery raid force is vulnerable to detection by forward enemy forces. To attack high value targets (which may be well within enemy territory), the artillery raid force may have to move around or very close to the enemy's forward forces. If the enemy has acquisition radar capabilities, the raid force could be vulnerable to counterbattery fires. Helicopter-borne artillery raid forces may be further exposed to enemy action due to the slow and staggered nature of insertion and withdrawal. These vulnerabilities increase the requirements for security and fire support for the raid force. The risk of detection can be reduced by conducting the raid under the cover of darkness and by selecting a firing position in defilade. The raid force must have an accurate survey capability, an accurate target location, and preferably "eyes-on-target" for rapid adjustment of fires and determination of target damage assessment.

METHODS

Artillery raids can be conducted by either sea-based units or by units deployed ashore. Sea-launched raids are categorized as amphibious. They have unique requirements and limitations, but offer some distinct advantages over shore-based raids. The alternative between amphibious and shore-based raids may be dictated to the commander by the deployment of his forces. Due to the extensive logistical task of transporting the raid force's howitzers, artillery raids are further categorized as either surface (transported by surface means) or helicopterborne. The four distinct types of artillery raids are described as *surface, helicopterborne, amphibious surface, and amphibious helicopterborne.* When considering the advantages and disadvantages of each type of raid, the commander's METT-T analysis will indicate which is the appropriate method. Advantages and disadvantages inherent in each type of raid are shown on page G-4.

METT-T ANALYSIS

Thorough and detailed METT-T analysis is an integral step in select-ing an artillery raid method and in the initiation of raid planning. METT-T analysis will largely dictate the general size, composition, and task organization of the raid force, its method of insertion, possible firing positions, time of attack, type and amount of ammunition to be fired, and security and support required.

Missions

Artillery raid forces can be given various missions depending on the commander's intent. Commander's intent, in concert with analysis of the threat, will in turn influence selecting the firing position, the insertion method, size, composition, and task organization of the raid force, and support and security required.

Destruction: fire delivered for the sole purpose of destroying material objects.

Neutralization: fire which is delivered to render the target ineffective or unusable.

METHOD	ADVANTAGES	DISADVANTAGES
Surface	Simple firing position criteria. Movement more discreet. Mobility of ammunition. Rapid occupation of firing positions. Rapid withdrawal from firing positions. Can be conducted without external support	Slower movement to/from firing position. Movement depends on roads/trafficability. Route may expose raid force to enemy ground forces. More ground security may be required Larger force required
Helicopterborne	Greater range. Larger area from which to select firing position. Rapid movement to/from firing position. Maneuver to avoid enemy/danger areas.	Complex LZ/firing position criteria. Staggered/slow occupation of and withdrawal from firing position. Extensive lift required for equipment/ammunition. Depend on weather/visibility. Vulnerable to antiaircraft artillery/small arms fire. Noise. Support for helicopter operations: - Escort helicopters - Loiter areas. - Forward aiming refueling point(s). - HST required at both departure and firing position.

		- Terminal guidance. - Helicopter routing.
Amphibious Surface (Assumes) use of LCACs)	Projection from sea; lodgement/logistics ashore not required. Rapid movement to/from firing position. Maneuver to avoid enemy/danger area. Greater lift capability.	Complex criteria for firing position/beach landing site. Depend on sea/surf/beach conditions. Slow debarkation from and reembarkation of LCACs. Noise. Complexity of ship-to-shore movement.
Amphibious Helicopterborne	Projected from sea; lodgement/logistics ashore not required. Greater range. Larger area from which to select firing positions. Rapid movement to/from firing position. Maneuver to avoid enemy/danger areas.	Complex firing position/LZ criteria. Staggered/slow occupation of and withdrawal from firing position. Extensive lift required for equipment/ammunition. Depend on weather/visibility. Vulnerable to antiaircraft artillery/small arms fire. Noise. Support for helicopter operations: - Escort helicopters. - Loiter areas. - Forward arming refueling point(s). - HST required at both departure and firing position. - Terminal guidance. - Helicopter routing. - Equipment access to flight deck. - Flight deck crew capability. - Time required to load force, assemble assemble waves, refuel helicopter.

Suppression: **degrading the performance of personnel or weapon systems to limit their ability to perform their mission for a specific time. This mission would generally be conducted as part of a combined arms raid or other larger operation.**

Harassment: reducing the enemy's combat effectiveness by disturbing the rest of his troops, lowering his morale, and curtailing his movement.

Diversion: drawing the attention and forces of the enemy from the principal operation either by contributing to ambiguity or deception.

The Enemy

Since an artillery raid is conducted at a standoff distance from the target, analysis of the threat to an artillery raid force is somewhat unique. The importance of enemy detection and counterfire capabilities are magnified. Route and firing position security is also very important due to artillery's vulnerability to air and ground attack. The threat will influence selecting the firing position, the method of insertion into the firing position, and the degree of security required enroute and in position. The risk of exposing critical artillery assets to enemy action must be balanced against the nature of the target(s) to be attacked.

Terrain

Availability of suitable LZs and beach landing sites that also meet the criteria required for firing positions will indicate which insertion method is feasible. The potential for surface movement will be based largely on the trafficability to, in, and from potential firing positions. Given the importance of rapid withdrawal, artillery raid forces cannot be hindered by poor trafficability in and around the firing position. These factors must be considered in selecting the best insertion method. Analysis of the terrain may also reveal a secure firing position in defilade that will significantly reduce the risk of detection and enemy action.

Troops

The capability and deployment of artillery units available for an artillery raid will influence many other decisions, particularly the best raid method. Available adequate assault support aircraft and landing craft will dictate whether helicopterborne or amphibious alternatives for raid force insertion are feasible. Analysis of unit training readiness will indicate what types of raids the force can conduct.

Timing

The cover of darkness may reduce the threat of detection and contribute to the effect of the raid on the enemy. However, benefits must be weighed against the inherent intricacies of night operations. Training and rehearsals are essential for a unit to overcome unique night obstacles.

Decisionmaking

Upon completion of this analysis, the commander will be able to make several decisions, including--

- If an artillery raid is appropriate.

- The method of insertion.

- The weapon (M198/M101A1).

- The unit to provide the raid force nucleus.

- What security is required.

- The raid forces's task organization.

- The type and amount of ammunition.

- The time of attack.

- Tentative firing positions.

- Tentative LZs/beach landing sites.

- Tentative routes to/from firing positions.

STANDING OPERATING PROCEDURES

Applying SOPs is critical in rapidly planning and preparing for a raid. Artillery units that can be expected to conduct raid operations should use operational and training experiences to develop an SOP that can serve as a basis for forming, training, and preparing a raid force. When tasked to conduct an artillery raid, commanders should use SOPs as a point of departure to ensure that specific, detailed planning considerations are addressed. SOPs should also detail who does what, how they do it, and in what specific sequence.

TASK ORGANIZATION

Personnel and equipment of the raid force should be limited to only those necessary to successfully complete the mission. Force size and structure will be significantly influenced by its planned insertion method and the security required. The force should be organized into these standard raid force elements: reconnaissance/security, assault, support, and command.

Reconnaissance/Security Element

This element may resemble the traditional artillery advance party. It may be formed from within the artillery raid unit or from a combi-nation of artillery and other personnel appropriate for the method of insertion and the threat. A FAC will normally be added to coordinate with supporting aircraft and direct any assault support and CAS.

For helicopterborne raids, a helicopter support team must be included. The reconnaissance/security element will normally be inserted in advance of the assault element to identify or select a specific firing position and/or howitzer positions; determine suitability and security of the site; secure the firing position/LZ/beach landing site; provide terminal guidance; and to support the raid force's arrival. It will normally remain in the firing position until the mission has been fired and the assault element has departed to cover the withdrawal.

Depending on the tactical situation, mission, and method of insertion, reconnaissance and security teams may be employed separately. In surface and amphibious surface raids, security teams may accompany the assault element throughout movement. The size of the reconnaissance/security element may be limited in helicopterborne, amphibious helicopterborne, and amphibious surface raids to reduce the insertion time and amount of lift required for the raid force.

Heavier forces; e.g., light armored vehicles and weapons-mounted, high mobility multipurpose wheeled vehicle (HMMWVs) can be incorpo-rated into the reconnaissance/security elements of surface raids without significantly slowing or hindering the raid force.

Assault Element

The assault element consists of the howitzers, other essential equipment, and the personnel necessary to operate them. The assault element may be required to provide its own security or may be accompanied by a portion or all of the security element in the manner of a traditional artillery main body.

Command Element

When required, a separate command element may be employed to provide command and control, communications, meteorological, and survey support to the raid force.

Support Element

Other artillery firing units or weapons-locating radars can serve as a support element and be employed in a supporting role. Employment of a second firing battery to the rear of the assault element to provide counterfires in support of the withdrawal of the raid force is an example of an artillery raid support element.

COMMUNICATIONS

To reduce the signature of the raid force, artillery raids should be conducted under the most restrictive EMCON that still enables effective control and coordination. Reliance on voice communications within the raid force is generally ineffective and can compromise noise discipline. Wire is not suitable for most raid force missions. Short-range radios (such as the AN/PRC-68) for internal communications are effective in maintaining noise discipline and limiting electronic emissions. The electronic signature of the force can also be limited by using execution checklists, brevity codes, and satellite communications or directional antennas for critical external communications.

FIRE SUPPORT

Fires must be planned to support employment of the raid force, particularly during insertion, position preparation, and the withdrawal. Scheduled and on-call fire support may be critical in protecting the artillery raid force and can contribute to the effect of the raid by attacking lucrative, high value targets exposed as a result of the raid. All fire support means including offensive air support, artillery, naval surface fire support, electronic warfare systems, and mortars must be considered. Electronic warfare assets (such as an EA-6B) can be extremely effective in protecting the raid force by jamming counterbattery radars just before and during target engagement. CAS can be effective if on-station ready to attack targets of opportunity (such as counterbattery units, counterfire systems, or ground units) identified during the raid. Another indirect fire unit employed as a support element positioned in trace of the assault element can effectively support the raid force's withdrawal or attack targets of opportunity. FACs, spotters, observers, and/or weapons locating radars must be incorporated into the appropriate raid force element.

SURVEY

There are numerous suitable methods of establishing survey in the firing position and laying the howitzers. The position and azimuth determining system (PADS) can be carried forward by the raid force to establish location and direction. GPS are more mobile and can quickly provide location. When using GPS, direction can be determined from astronomical observation, a north-finding module, magnetic means, or by transfer from another position by simultaneous observation. The method of establishing survey and laying the howitzers must be fast, accurate, and consistent with the lift available to the raid force. It should also not degrade noise and light discipline.

COMPUTATION OF FIRING DATA

Howitzer firing data for raids will normally be determined using the handheld backup computer system (BUCS). BUCS is the most rapid, flexible, simple, and mobile means of computing firing data. If the firing position, target, and meteorological conditions can be accurately identified in advance, firing data can be computed and provided to the howitzer section chiefs before arriving in the firing position. Although this will not be possible in most situations, known elements to compute firing data (such as target location) should be entered into the BUCS before arriving in the firing position. Only those personnel needed to enter the final elements and compute and verify firing data should accompany the raid force.

AMMUNITION

The raid force should limit ammunition to that required to achieve desired effects on the target and meet any other mission requirement. Consistent with safety restraints, ammunition should be prepared for firing in advance to limit time in the firing position.

REHEARSALS

Preparation of personnel should include rehearsals to the extent practical. Since an artillery raid typically requires various supporting and security forces external to artillery units, rehearsals are critical in bringing the artillery raid force together and ensuring common understanding. Rehearsals also provide the opportunity to exercise these integrated forces and test the complex techniques inherent in discreetly moving this size force. Safety restraints may limit the extent of any rehearsal.

EXECUTION

Execution of the artillery raid must be well-coordinated. An execution checklist or timeline should be used to coordinate activities of the raid force and supporting forces. Go/no-go criteria should be established so the required conditions for success have been met before the raid force is unnecessarily exposed. Examples of these conditions include accurate target location, CAS is on-station, the minimum number of howitzers arrive safely in position, and positive communications are established between the raid force and the appropriate commander.

Target Engagement

Target engagement can be planned for initiation based on--

- When the raid force is in position and ready.

- At a specific pre-designated and coordinated time.

- An enemy event or action; e.g., raid force assumes "ambush" position and fires on a high value target when it appears or exposes itself.

- Command of higher authority.

Emergency Withdrawal

The plan should also specify conditions for emergency withdrawal; e.g., cease firing and withdraw if the raid force receives indirect fire. Artillery raid withdrawals can be triggered by--

- Time; e.g., cease firing at 0210 and withdraw.

- Limiting the duration of firing and thus the exposure to acquisition radar; e.g., cease firing within 3 minutes after engagement and withdraw.

- Firing the requisite amount of ammunition to achieve the desired effects; e.g., cease firing after firing 32 rounds dual-purpose improved conventional munitions and withdraw.

- A combination of these three.

Appendix H
Glossary

Acronyms and Abbreviations

AAV	assault amphibious vehicle
ACE	aviation combat element
AMC	air mission commander
ARG	amphibious ready group
Bn	battalion
BUCS	backup computer system
CAS	close air support
CATF	commander, amphibious task force
CLF	commander, landing force
CO	Commanding Officer
CRRC	combat rubber raiding craft
E&E	evasion & escape
EEI	essential elements of information
EMCON	emission control
FAC	forward air controller
FMFM	Fleet Marine Force manual
GCE	ground combat element
GPS	global positioning system
HMMWV	high mobility multipurpose wheeled vehicle
hp	horsepower
ITG	initial terminal guidance
LCAC	landing craft air cushion
LCU	landing craft, utility
loran	long-range aid to navigation

LPD .... landing platform dock
LRD .... launch and recovery device
LRDG .... long-range desert group
LSD .... landing ship dock
LST .... landing ship tank
LZ .... landing zone

MAGTF .... Marine Air-Ground Task Force
METT-T .... mission, enemy, terrain and weather, troops and support available—time available
MEU .... Marine expeditionary unit

NAVAIDS .... navigational aids
NCO .... noncommissioned officer
NEO .... noncombatant evacuation operation

OIR .... other intelligence requirements

PADS .... position and azimuth determining system
PIR .... priority intelligence requirements

RHIB .... rigid hull inflatible boat
RRC .... rigid raiding craft
RTO .... radio/transmitter operator

SAS .... Special Air Service
SMAW .... shoulder-launched, multipurpose assault weapon
SOP .... standing operating procedure

TAC .... tactical air controller
TOW .... tube launched, optically tracked, wire command link, guided missile system

ZANLA .... Zimbabwe African National Liberation Army

Index

A

B

C

D

E

F

N

O

S

T

U

Made in the USA
Monee, IL
08 May 2026

49751718R00095